Workshop on Processing Physic-Chemistry Advanced 2016 (WPPCA 2016)

AA006742

Journal of Physics: Conference Series Volume 743

Bucaramanga, Columbia
4 – 8 April 2016

Editor:

Ely Dannier V. Nino

ISBN: 978-1-5108-2916-9
ISSN: 1742-6588

Printed from e-media with permission by:

Curran Associates, Inc.
57 Morehouse Lane
Red Hook, NY 12571

proceedings.com

Some format issues inherent in the e-media version may also appear in this print version.

Copyright© (2016) by the Institute of Physics
All rights reserved. The material featured in this book is subject to
IOP copyright protection, unless otherwise indicated.

Printed by Curran Associates, Inc. (2016)

For permission requests, please contact the Institute of Physics
at the address below.

Institute of Physics
Dirac House, Temple Back
Bristol BS1 6BE UK

Phone: 44 1 17 929 7481
Fax: 44 1 17 920 0979

techtracking@iop.org

Additional copies of this publication are available from:

Curran Associates, Inc.
57 Morehouse Lane
Red Hook, NY 12571 USA
Phone: 845-758-0400
Fax: 845-758-2633
Email: curran@proceedings.com
Web: www.proceedings.com

Table of contents

Volume 743

Workshop on Processing Physic-Chemistry Advanced (WPPCA)
4–8 April 2016, Bucaramanga, Santander, Colombia

Accepted papers received: 9 August 2016
Published online: 31 August 2016

Preface

Papers

Journal of Physics: Conference Series **743** (2016) 011001 doi:10.1088/1742-6596/743/1/011001

Preface

In the present volume of Journal of Physics: Conference Series we publish the proceedings of the "2nd Workshop on Processing Physic-Chemistry advanced (WPPCA)", that was held from, April 4-8, 2016, at the Universidad Industrial de Santander (UIS), Bucaramanga, Colombia. The proceedings consist of 17 contributions that were presented as plenary talks at the event. The abstracts of all participants contributions were published in the Abstract Book with ISSN 2500-8420. The website of the symposium is available at http://www.sc3.uis.edu.co/workshop-en-procesamiento-fisico-quimico-avanzado-wppca/.

The scientific program of the 2nd WPPCA consisted of 12 Magisterial Conferences, 28 Poster Presentations and 2 Courses with the participation of undergraduate and graduate students, professors, researchers and entrepreneurs from Colombia, Spain, Unite States of America, Mexico and Chile. Moreover, the 2nd WPPCA allowed to establish a shared culture of the research and innovation that enriches the area of the processing physical-chemistry of the materials and the industrial applications.

All papers in these Proceedings refer to one from the following topics: Semiconductors, Superconductivity, Nanostructure Materials and Modelling, Simulation and Diagnostics. The editor hopes that those interested in the area of the science of materials can to enjoy this reading, that reflects a wide variety of current issues.

On behalf of the organizing committee of the 2nd WPPCA, we are extremely thankful to all authors for providing their valuable contributions for these Proceedings as well as the reviewers for their constructive recommendations and criticism aiding to improve the presented articles. Besides, especially we appreciate the great support provided by the Sponsors and Partners.

The Editor

Ely Dannier V. Niño

Content from this work may be used under the terms of the Creative Commons Attribution 3.0 licence. Any further distribution of this work must maintain attribution to the author(s) and the title of the work, journal citation and DOI.
Published under licence by IOP Publishing Ltd

Journal of Physics: Conference Series **743** (2016) 011001 doi:10.1088/1742-6596/743/1/011001

Organizing Committee
Carlos Leonardo Beltrán Rios (Chairman)
Carlos Jaime Barrios Hernández
Harold Paredes Gutiérrez
Ely Dannier V. Niño

National Scientific Committee
Rafael Julian González Hernández
Jairo Arbey Rodríguez Martínez
Carlos Jaime Barrios Hernández
Carlos Leonardo Beltrán Ríos
Jesús David González Acosta
Harold Paredes Gutiérrez
Javier Antonio Montoya
José José Barba Ortega
Ederley Vélez Ortiz
Julián David Correa
José Sierra Ortega

International Scientific Committee
Julibeth Martínez, Unite State of America
Eduardo Ariel Menendez Proupin, Chile
Felipe Gonzalez Cataldo, Chile
Antonio Rivera de Mena, Spain

Sponsors
- ➢ Universidad Industrial de Santander

Partners
- ✓ Grupo de Investigación en Física Computacional en Materia Condensada (FICOMACO)
- ✓ Foundation of Researchers in Science and Technology of Materials (FORISTOM)
- ✓ Grupo de Investigación en Física y Tecnología del Plasma y Corrosión (FITEK)
- ✓ Grupo de Investigación en Física Mesoscópica
- ✓ Centro de Materiales y Nanociencias (CMN)
- ✓ J & G Soluciones Industriales
- ✓ Canaán Multiservicios
- ✓ S & S Ingnieria S.A.S

Journal of Physics: Conference Series **743** (2016) 011001 doi:10.1088/1742-6596/743/1/011001

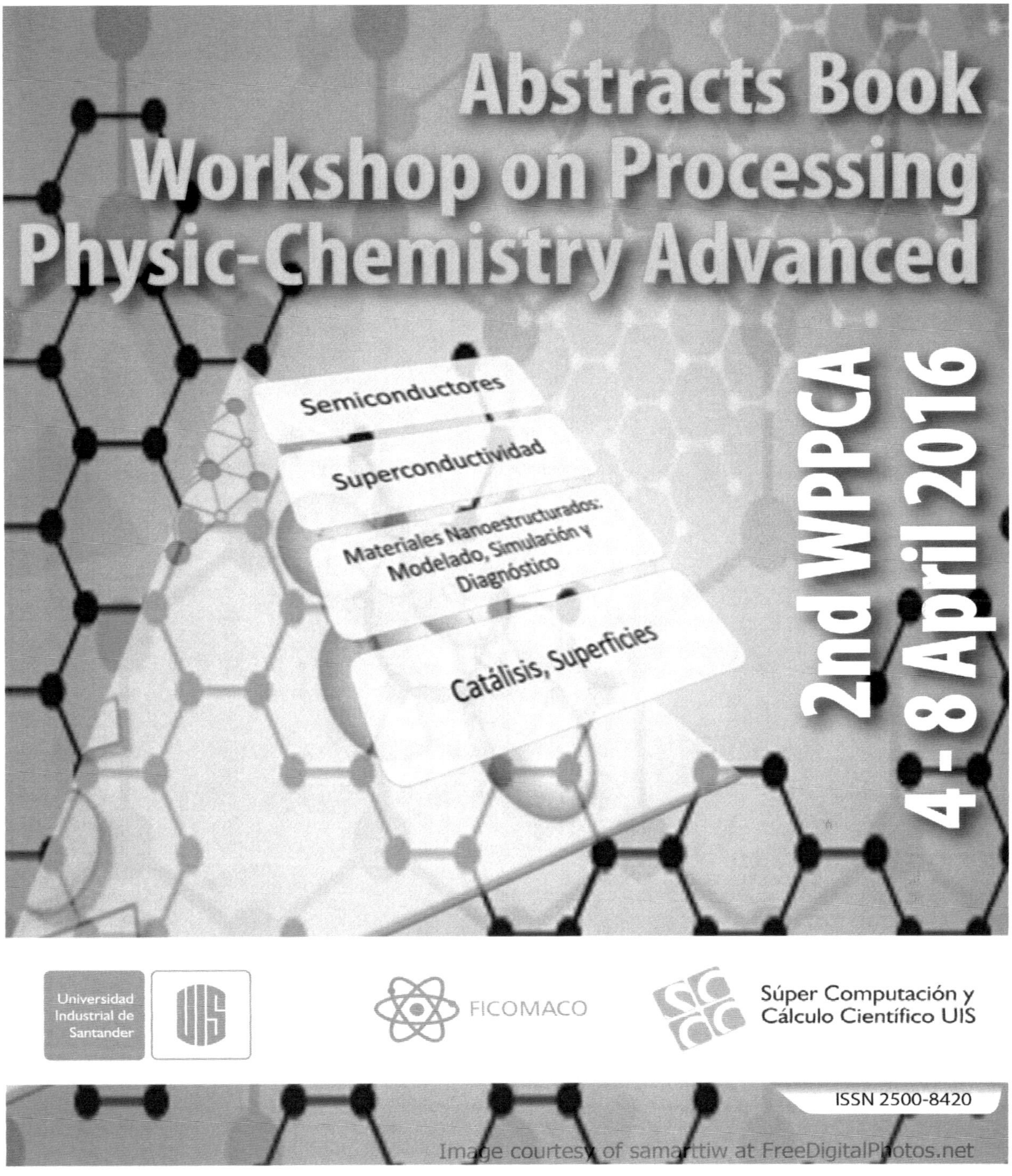

Journal of Physics: Conference Series **743** (2016) 011001 doi:10.1088/1742-6596/743/1/011001

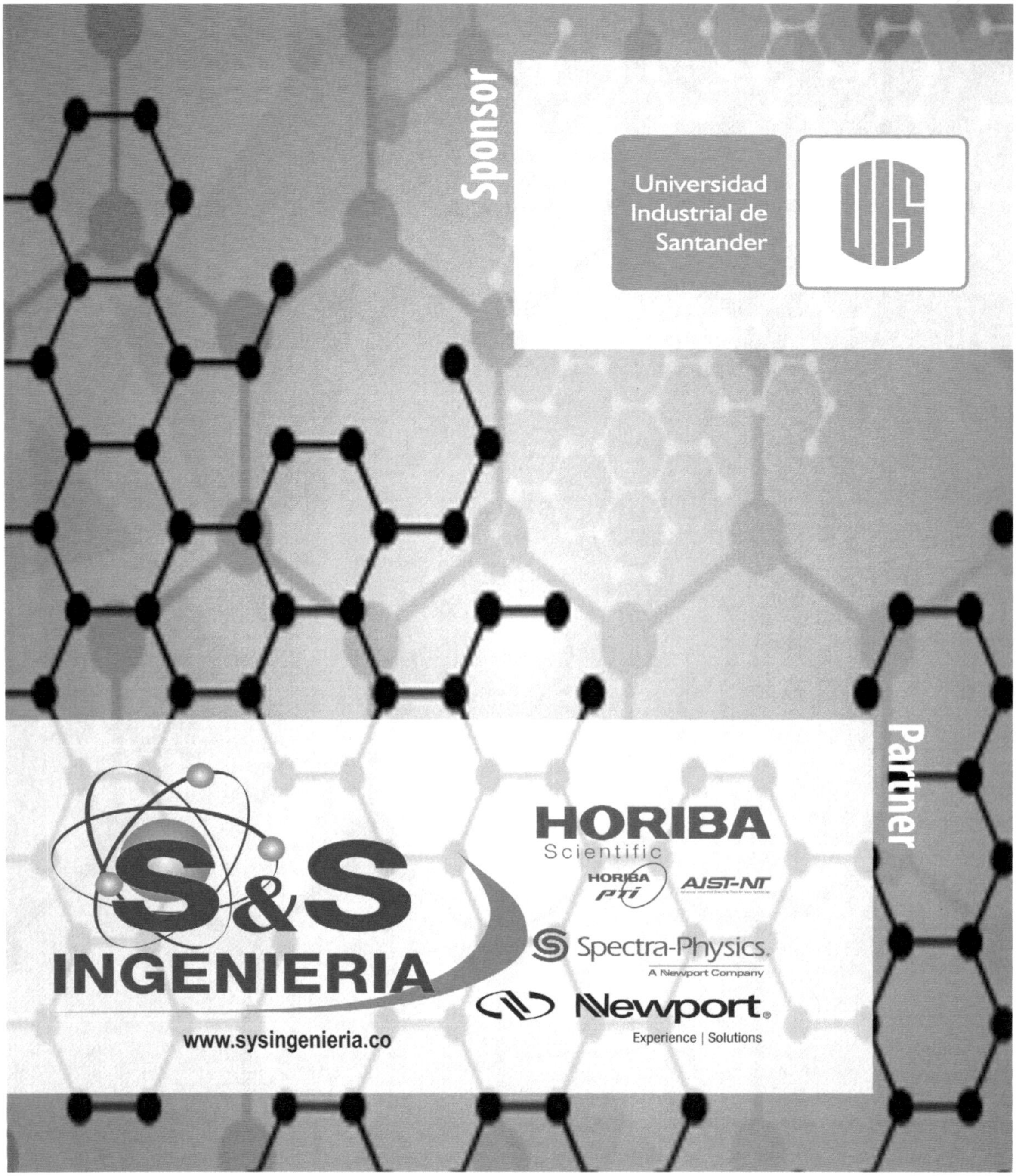

Peer review statement

All papers published in this volume of *Journal of Physics: Conference Series* have been peer reviewed through processes administered by the proceedings Editors. Reviews were conducted by expert referees to the professional and scientific standards expected of a proceedings journal published by IOP Publishing.

Content from this work may be used under the terms of the Creative Commons Attribution 3.0 licence. Any further distribution of this work must maintain attribution to the author(s) and the title of the work, journal citation and DOI.
Published under licence by IOP Publishing Ltd

Ab-initio structural search in solid oxygen at high pressure: from zero to finite temperature

B H Cogollo-Olivo[1] **and J A Montoya**[1,2]
[1] Universidad de Cartagena, Cartagena de Indias, Bolívar, Colombia.
[2] The Abdus Salam ICTP, Trieste, Italy.

E-mail: bcogolloo@unicartagena.edu.co

Abstract. The crystal structure of solid oxygen in the terapascal (TPa) regime has been investigated with Density Functional Theory and the Random Search algorithm at zero temperature. We also considered the effect of the entropy at finite temperatures using the Quasi-Harmonic Approximation, and we found that the regime of stability of solid oxygen differs strongly from the results predicted at zero temperature. Finally, we provide some insights of oxygen as a chalcogen element.

1. Introduction

Oxygen is a very interesting object of study: for instance, among common diatomic molecules (H2, N2, O2, CO, F2, etc.), oxygen is the only element that can carry a magnetic moment [1]. Aside from its abundance in the Universe, its importance for many fundamental biological functions, or its role in the absorption of solar radiation inside Earth's atmosphere, oxygen exhibits an unique combination of characteristics that make it a physical object of great interest for the scientific community. Nowadays, only molecular phases of solid oxygen have been found via experimental data, even at the highest pressures currently attainable in hydrostatic compression experiments. Up to this date, six solid phases of oxygen have been established unambiguously: half of them (α, β and γ) exist under equilibrium vapour pressure, and the other three (δ, ϵ and ζ) are obtained in the high pressure regime. Nevertheless, in the atmospheres of giant planets, inside and outside our solar system, non-molecular oxygen could be much more abundant than the molecular phases due to the extreme pressure and temperature conditions that are present. This hypothesis has motivated several theoretical and experimental studies aiming to get a better understanding of oxygen under those conditions, so that we can have a better comprehension of the processes that occur in planetary interiors.

Upon increasing pressure at low temperature, a first transformation takes place from α-oxygen to the anti-ferromagnetic δ-phase at approximately 6GPa [2, 3, 4, 5]. δ-phase is orthorhombic and it belongs to the space group Fmmm. In [5] another oxygen phase was described, which appeared at 10GPa and 300K and exhibited a visual change of the sample colour from light orange to dark red; this type of change is a particularity of solid oxygen. This new oxygen form was called ϵ-phase.

Several studies were carried out in order to define the structure of the ϵ-phase: optical [6, 7] and X-Ray powder diffraction studies [8, 9, 10] have demonstrated that the ϵ-phase retains a layered structure with parallel alignment of molecules that is typical for all known oxygen structures except for γ-oxygen. Gorelli *et al.* [11] suggested that ϵ-oxygen's most basic unit is not an O_2 molecule, but a cluster of molecules. This assumption was confirmed by X-Ray diffraction [12, 13]: ϵ-oxygen is an arrangement of groups of four O_2 molecules that form a monoclinic C2/m structure. The E-phase has been widely

Content from this work may be used under the terms of the Creative Commons Attribution 3.0 licence. Any further distribution of this work must maintain attribution to the author(s) and the title of the work, journal citation and DOI.
Published under licence by IOP Publishing Ltd

studied using different DFT functionals and it has been demonstrated that for pressures above 30GPa, DFT methods are in very good agreement with experimental results [14]. A recent study [15] has proposed a new phase diagram for solid oxygen, where the single E-phase should be replaced with two phases $\epsilon_1 - O_2$ and $\epsilon_2 - O_2$: the first one is a local singlet spin 1 liquid, while the second one is a regular, Peierls band insulator separated by a first order phase transition near 20GPa.

A room temperature X-Ray study revealed that the E-phase remains stable up to 96GPa, when it suffers another transformation and becomes metallic in the ζ-phase [8]. This transition, from insulator to a metallic phase is unique, but to date remains unexplained. The lack of definitive experimental information on the atomic positions inside the unit cell in the ζ phase has made that the $\epsilon - \zeta$ phase transition is still a matter of discussion. But this phase is not only metallic: a theoretical study predicted that at 0.6K and 100GPa, the solid oxygen turns into a superconductor [16], and this result was later confirmed experimentally [17].

The scenario for even higher pressures is also intriguing, given the uncertainty regarding the question if the pressure applied to solid oxygen will lead to a molecular dissociation or further association: theoretical studies now suggest that even in the terapascal (TPa) regime oxygen remains molecular [18], in contrast with the fact that other diatomic molecules have suffered a molecular dissociation at much lower pressures [19-23].

In the present study we started by extrapolating the ζ-phase's equation of state to find a suitable density for solid oxygen around 350GPa in order to start increasing from there the external pressure that is applied on this system, up to 10TPa. At each pressure point, we performed a structural search looking for post-ζ-oxygen phases, in particular non-molecular ones. Our results agree with previous theoretical studies, which had already predicted non-molecular phases to appear only above 2000GPa (2TPa) [18], however, the range of pressures covered in our study offers more detail than previous studies and can still shed some new light about the nature of the molecular oxygen bonding at high pressures. In addition, we have performed calculations of a sizable contribution to the free energy of the system that comes from the entropy at finite temperatures, originated in the vibration degrees of freedom of the system. Our calculations at this level suggest that the regime of stability of the non-molecular forms, at temperatures of relevance for planetary sciences, differs strongly from the results predicted at zero temperature by [18].

2. Methodology

After deciding on an initial cell-volume to target a pressure in the range from 0.35 to 10TPa, triclinic structures of different sizes were randomly generated which had up to four atoms per unit cell. We started by generating randomly a set of cell parameters and then re-scaled them to fit the estimated volume at the target pressure [24]. Before re-scaling and in order to avoid needle-like or other very distorted shapes of the starting unit cell, the crystal lattices were generated by selecting three random numbers between 0.5 and 1.5 as the lengths of the cell-vectors and three random angles between 40° and 140°. Finally, we generated randomly up to four atomic positions inside the cell, filtering out cases with severe overlap between atoms. Following this recipe we generated a thousand samples for each pressure in this study.

Each sample was then relaxed until its atomic positions and cell parameters reached a nearby local minimum inside the free-energy landscape by following the gradient. It has been shown several times that by using this method, the set stable structures at different pressures, which are the ones with the lowest enthalpy, can be found and characterized in a reliable manner [18, 19, 25-31]. The force and energy calculations needed for the structural relaxations were done ab-initio, using the pw.x code from the Quantum ESPRESSO package [32]. We used the Perdew-Burke-Ernzerhof (PBE) generalized gradient (GGA) exchange- correlation functional [33], with ultra-soft pseudo-potentials and the Brillouin-Zone integrations were approximated with a converged 8×8×8 k-points grid using the method of Monkhorst and Pack [34]. A plane-wave kinetic energy cutoff of 60Ry was selected, which is high but needed in order to calculate the stress on the cell accurately.

Journal of Physics: Conference Series **743** (2016) 012001 doi:10.1088/1742-6596/743/1/012001

Given the unbiased nature of the structural search method being used, we often find ourselves starting from rather far-from-equilibrium configurations of the cell parameters and the atomic positions. Therefore, as a norm, we first ran a loosely converged calculation aimed to relax the system close enough to a nearby enthalpy minimum. Once we got this approximated structure, we performed a second more strict relaxation, using the k-point and cutoff parameters from the convergence tests that were mentioned before (in the tests, total energy differences had been converged to better that 5meV per atom).

3. Results and discussion

3.1. Molecular solid oxygen

Every set of a thousand samples converged into a reduced number of local minima and, according to the high number of repetitions observed, we can say with confidence that there are no other energy minima available at each energy landscape studied. Since we are performing a structural search for pressures that are higher than zero, our comparison between structures was not done with the total energy, but with their enthalpies instead.

Once the global enthalpy was determined, we selected the best crystal structure, i.e. the one with the lowest enthalpy among all, at each studied pressure. Since we assume all our cells to be triclinic, the package FINDSYM [35] from the ISOTROPY Software Suite was used in order to find an equivalent conventional cell with higher symmetry that made further analysis easier. Over the entire range of pressures studied, we found three different symmetry groups for the unit cells with two oxygens: triclinic (P-1), monoclinic (C2/m) and orthorhombic (Cmcm); for the unit cells with four oxygens we found two kinds of orthorhombic structures (Cmma and Cmcm). At pressures above 0.55TPa for the small cell and 1.5TPa for the big one, the most stable structure was a dense zigzag-chains Cmcm phase. At 0.50TPa, the bonds of the O_2 system are not completely symmetric to one another: we observed that from 0.40TPa first neighbour distance increased significantly, while the second decreased. We found that when pressure was increased to 0.55TPa, the two shortest interatomic distances became symmetrical and the oxygen coordination went from one to two (see Figure 1(a)), therefore, the first neighbour and the second neighbour distances converge at 0.55TPa. A similar transition was found at 1.30TPa for the O_4 unit-cell (see Figure 1(b)), in both cases much earlier than predicted by other studies [18].

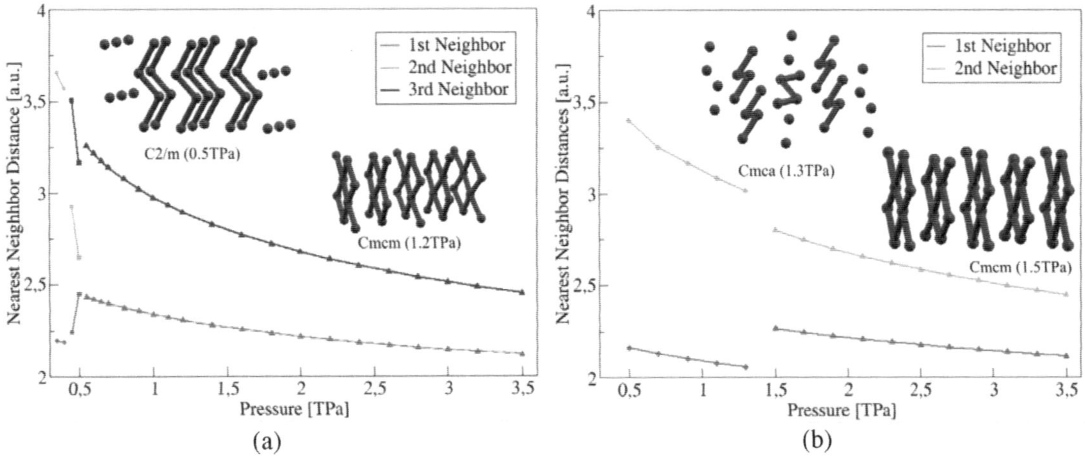

Figure 1. Evolution of the closest oxygen distances while pressure increases. (a) O_2 system structures: P-1 (circles), C2/m (squares) and Cmcm (triangles). (b) O_4 system structures: Cmma (diamonds) and Cmcm (triangles).

For both unit-cell sizes the stable structure at higher pressures is Cmcm, however for the O_4 unit-cell this symmetry group shows up at higher pressures than for the O_2 unit-cell. Due to the smaller amount of degrees of freedom available to the O_2 system, the only arrangement that it can achieve at 0.55TPa is Cmcm. The O_4 unit cell instead, having more atoms available to explore possible configurations, remains with the Cmma structure, which seems to be more favourable energetically, until it reaches to 1.50TPa. By analysing the enthalpy per atom from 0.35 to 1.50TPa, the enthalpy for the O_4 system is lower than for the O_2 unit-cell, but from 1TPa the enthalpy per atom of the O_2 is better by a few hundredth electron-volts.

3.2. Molecular solid oxygen at finite temperature

At extreme conditions of pressure and temperature, similar to those found at planetary interiors, the thermodynamical behaviour of the materials plays a significant role in the description of their physical properties. Accurate calculations regarding the effects related to lattice dynamics can be done by using the Quasi-Harmonic Approximation (QHA) [36, 37]. In the QHA, the contribution to the vibrational part of the free energy is determined by:

$$F_{vib} = \frac{1}{2}\sum_{qs} \hbar w(qs) + k_B T \sum_{qs} \ln[1 - exp(-\hbar w(qs)/k_B T)] \quad (1)$$

And the entropy can be derived as: $S = -(\partial F / \partial T)_V$. We performed finite temperature calculations from T=0K to T=2500K to study the phase-transition boundary from Cmcm to Fmmm that occurs around 9.3TPa at zero temperature, according to [18]. The most remarkable feature of our findings was that even without the zero-point energy contribution, there was a shift to the left of the non-molecular phases transition point. We found that while the temperature is increasing, the merging from Cmcm to Fmmm occurs at lower pressures due to purely to entropic contributions (see Figure 2).

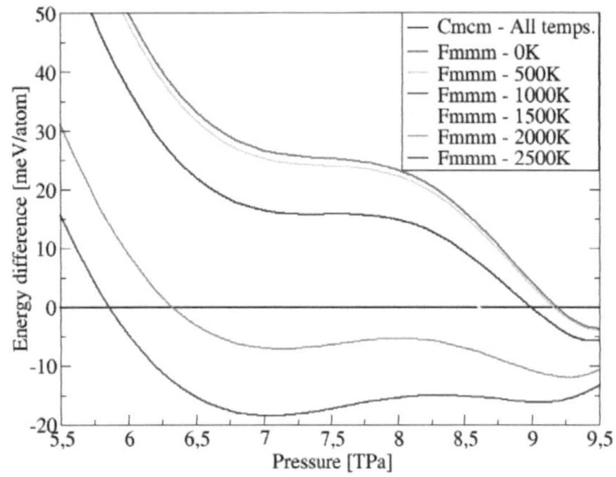

Figure 2. Enthalpy difference for the Fmmm structure at finite temperature from 0 to 2500K. Cmcm structure is the reference.

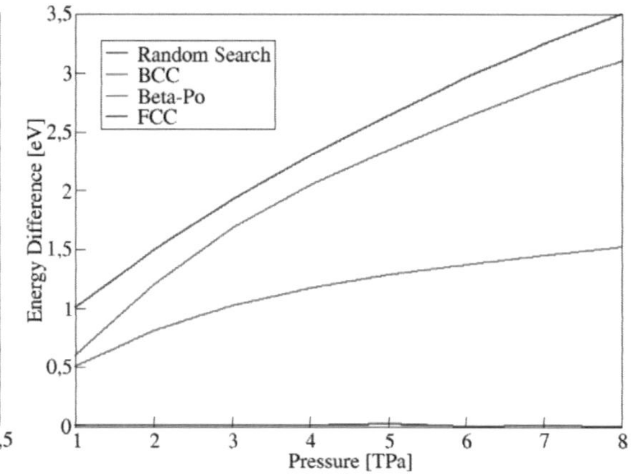

Figure 3. Enthalpy difference for several monoatomic oxygen structures from 1TPa to 8TPa. Fmmm structure is the reference.

3.3. Monoatomic oxygen (special case)

Up to 9TPa solid oxygen remains molecular while other chalcogen elements (group 16 of the periodic table) take a monoatomic rhombohedral β-Po structure after being compressed [38-40]. Going beyond, when further compression is applied on Se and Te, a BCC structure is formed after their β-Po-type structure [40-42], however a layered monoatomic crystalline phase for oxygen has been predicted to occur only above 9TPa [18]. Given that oxygen exhibits several different structures [18], such as chain-like I41/acd at 2TPa, a Cmcm zigzag structure at 3TPa, and finally a monoatomic Fmmm structure at 9.3TPa, we decided to complete the picture of oxygen as a chalcogen by making an structural search for monoatomic oxygen up to 8TPa, and then adding a comparison between those structures that we found and the β-Po, BCC and FCC (see Figure3).

For this study we used a slightly different approach in order to make our findings comparable with the results reported in [18]. We used Perdew-Burke-Ernzerhof (PBE) generalized gradient (GGA) exchange-correlation functional, a k-point grid of 20×20×20 due to the very small cells involved and a plane wave kinetic cutoff energy of 60Ry. Also, for the most stable structures from the random search, we performed very well converged calculations using hard Norm-Conserving pseudopotentials and nonlinear core corrected projector augmented-wave (PAW), with the same exchange-correlation functional.

We found that for pressures above 1TPa, the most stable structure is Fmmm. In fact, it is evident that for higher pressures the enthalpy difference increases for all the other structures that were analysed. Although they are metastable, the BBC structure was observed as a post- β-Po phase same as for Te and Se [40-42]. We made calculations that are not shown in Figure 3 in order to establish the pressure at which the transition from β-Po to BCC takes place, and we found that it is slightly above of 0.9TPa. Even if this pressure is much higher than the ones determined for Te and Se, it is clear that oxygen follows the same trend and therefore it has something in common at some level with other chalcogen elements.

4. Conclusions

Due to technical limitations, hydrostatic compression experiments can reach pressures up to 300GPa, value that is close to the pressure in the Earth's inner core. At that pressure regime, oxygen keeps its molecular form. Using computational approaches, we were able to study this system at pressures that today are impossible to obtain in a laboratory.

The dramatic changes found for molecular solid oxygen at pressures below 1.50TPa were also observed by Sun *et al.* [18], but we offered more detailed calculations that are necessary for a better understanding of the structural and physical changes that occur under those conditions. The structural search calculations were complemented with a tool for prediction of the thermal properties, such as QHA. The finite temperature calculations which include the vibrational entropy show that the pressures for molecular dissociation are not as high as those predicted by [18] at the temperatures of interest for planetary interiors.

Regarding the structural search calculations done for monoatomic oxygen, we observed that it is not like other chalcogen element. Our curves comparing the stability of the Fmmm structure found in [18] with other possible monoatomic structures observed in chalcogens, reveal that there is no point above 1TPa where any of those structures are more stable than Fmmm. Indeed, the fact that the energy difference between them actually grows with pressure, discards any possible resemblance of oxygen to other chalcogens, even going to higher pressures. It is however interesting that at least as metastable structures the β-Po oxygen is lower in enthalpy with respect to BCC until BCC takes over at around 0.9TPa, and this sequence at least is consistent with the behaviour found in Se and Te.

References

[1] Freiman Y A and Jodl H J 2004 *Phys. Rep.* **401** 1–288
[2] Schiferl D *et al.* 1983 *Acta Crystallographica Section B* **39** 153–157
[3] Gorelli F A *et al.* 2002 *Phys. Rev. B* **65** 172106
[4] Goncharenko I N *et al.* 2004 *Phys. Rev. Lett.* **93** 055502
[5] Nicol M *et al.* 1979 *Chemical Physics Letters* **68** 49–52
[6] Syassen K and Nicol M 1981 Solid O_2 near 298k: Raman and electronic spectra of β-and $\epsilon - O_2$ *Proc. Int. Symp. Phys. Sol. High Pressure* ed Schilling J S and Shelton R N (Amsterdam: North-Holland Publi. Co.) pp 33–38
[7] Nicol M and Syassen K 1983 *Phys. Rev. B* **28** 1201–1206
[8] Akahama Y *et al.* 1995 *Phys. Rev. Lett.* **74** 4690–4693
[9] Desgreniers S and Brister K 1996 Crystalline structure of the high density ϵ-phase of solid O_2 *High Pressure Science and Technology* ed Trzeciakowski W A (Singapore: World Scientific) pp 363–365

Journal of Physics: Conference Series **743** (2016) 012001 doi:10.1088/1742-6596/743/1/012001

[10] Weck G *et al.* 2002 *Phys. Rev. Lett.* **88** 035504
[11] Gorelli F A *et al.* 1999 *Phys. Rev. Lett.* **83** 4093
[12] Lundegaard L F *et al.* 2006 *Nature* **443** 201–204
[13] Fujihisa H *et al.* 2006 *Phys. Rev. Lett.* **97** 085503
[14] Bartolomei M *et al.* 2014 *Chemical Physics Letters* **592** 170–174
[15] Crespo Y *et al.* 2014 *Proceedings of the National Academy of Sciences* **111** 10427–10432
[16] Ma Y *et al.* 2007 *Phys. Rev. B* **76** 064101
[17] Weck G *et al.* 2009 *Phys. Rev. Lett.* **102** 255503
[18] Sun J *et al.* 2012 *Phys. Rev. Lett.* **108** 045503
[19] Pickard C J and Needs R J 2007 *Nat. Phys.* **3** 473–476
[20] Fujii Y *et al.* 1989 *Phys. Rev. Lett.* **63** 536–539
[21] Takemura K *et al.* 1980 *Phys. Rev. Lett.* **45** 1881–1884
[22] Eremets M I *et al.* 2001 *Nature* **411** 170–174
[23] Ma Y *et al.* 2009 *Phys. Rev. Lett.* **102** 065501
[24] Pickard C J and Needs R J 2011 *J. Phys. Condens. Matter* **23** 053201
[25] Pickard C J and Needs R J 2006 *Phys. Rev. Lett.* **97** 045504
[26] Pickard C J and Needs R J 2007 *J. Chem. Phys.* **127**
[27] Pickard C J and Needs R J 2007 *Phys. Rev. B* **74** 144114
[28] Pickard C J and Needs R J 2009 *J. Phys. Condens. Matter* **21** 452205
[29] Pickard C J and Needs R J 2009 *Phys. Rev. Lett.* **102** 125702
[30] Pickard C J and Needs R J 2009 *Phys. Rev. Lett.* **102** 146401
[31] Morris A J *et al.* 2008 *Phys. Rev. B* **78** 184102
[32] Giannozzi P *et al.* 2009 *J. Phys. Condens. Matter* **21** 395502
[33] Perdew J P *et al.* 1996 *Phys. Rev. Lett.* **77** 3865–3868
[34] Monkhorst H J and Pack J D 1976 *Phys. Rev. B* **13** 5188–5192
[35] Stokes H T and Hatch D M 2005 *J. Appl. Crystallogr.* **38** 237–238
[36] Leibfried G and Ludwig W 1961 *Solid State Phys.* **12** 275–444
[37] Baroni S *et al.* 2010 *Rev. Mineral. Geochem.* **71** 39–57
[38] DeSando R J and Lange R C 1966 *J. Inorg. Nucl. Chem.* **28** 1837–1846
[39] Jamieson J C and McWhan D B 1968 *J. Chem. Phys.* **43** 1149–1152
[40] Akahama Y *et al.* 1993 *Phys. Rev. B* **25** 17–33
[41] Hassan F E H *et al.* 2005 *Physica B: Condensed Matter* **363** 82–87
[42] Otani M and Suzuki N 2001 *Phys. Rev. B* **63** 104516

Ab-initio study on electronic and magnetic properties of (Ga,Co) co-doped ZnO

A González-García[1], V Mendoza-Estrada[1], W López-Pérez[1], C Pinilla-Castellanos[1] and R González-Hernández[1]
[1] Universidad del Norte, Barranquilla, Colombia.

E-mail: alvarogonzalez@uninorte.edu.co

Abstract. Using first-principles calculations based on density functional theory within GGA formalism, we have studied the electronic structure and magnetic properties of (Ga,Co) co-doped ZnO system. The effect of impurity distances on ferromagnetic and antiferromagnetic ground state in $Co_{0.056}Zn_{0.944}O$ has been studied. For the closest Co-Co distance, a ferromagnetic ground state with total magnetic moment of $\sim 3.00\mu_B$ per Co atom has been found. The electronic structure also displays a nearly halfmetallic order. Conversely, for the farthest Co-Co distance an antiferromagnetic ground state was found for $Co_{0.056}Zn_{0.944}O$. When Zn^{2+} ions are replaced by Ga ions in $Co_{0.056}Zn_{0.944}O$, the new (Ga,Co) co-doped ZnO system is more energetically stable. It has also been found that Ga-doping reduces the $Co_{0.056}Zn_{0.944}O$ band gap due to the sp-d exchange interactions, which is in good agreement with the experimental data. Moreover, the Ga-doping changes the nearly halmetallic order of $Co_{0.056}Zn_{0.944}O$ to metallic. Results also show that $Ga_{0.029}Co_{0.056}Zn_{0.915}O$ is still ferromagnetic with a total magnetic moment of $\sim 3.00\mu_B$ per Co atom. It was also found that the ferromagnetic ground state in (Ga,Co) co-doped ZnO vanishes as Ga concentration increases.

1. Introduction

ZnO is a direct, wide bandgap semiconductor which has been widely used for its potential applications in manufacturing blue/ultraviolet (UV) light-emitting diodes (LEDs) and transparent electronics devices [1]. During the last years there has been predicted room temperature ferromagnetism in ZnO-doped with transition metal ions (TM) both theoretically and experimentally [2, 3, 4]. The above makes TM-doped ZnO an attractive material for spintronics applications. Spintronics or spin-electronics is a multidisciplinary new field of physics and technology that aims to manipulate and control the spin degrees of freedom in solid state systems [5]. Pandey *et al.* found experimentally that ZnO:Ni films are ferromagnetic up to room temperature [4]. Authors stated that Ni-doped ZnO is an important candidate for transparent magnetic semiconductor. The ferromagnetic behaviour in Ni-doped II-VI and III-V semiconductors has also been studied by ab-initio calculations [2, 6]. Ferromagnetism has also been predicted for Cr-doped ZnO by ab-initio calculations [7]. Conversely, Spaldin found by DFT study that ferromagnetism is not strongly favoured in ZnO doped with either Co or Mn on the Zn site, unless additional dopants which provide carriers (holes) are also incorporated [8]. Therefore, doping ZnO with trivalent dopant could be efficient to improve the ferromagnetism in Co-doped ZnO. As a matter of fact, Jun *et al.* studied the structural, optical, electrical and magnetic properties of $Zn_{1-x}Co_xO$:Al prepared by sputtering [9]. Their results suggest that codoping Co-doped ZnO with trivalent Al atoms could not only improve its ferromagnetic properties but also spread

Content from this work may be used under the terms of the Creative Commons Attribution 3.0 licence. Any further distribution of this work must maintain attribution to the author(s) and the title of the work, journal citation and DOI.
Published under licence by IOP Publishing Ltd

Journal of Physics: Conference Series **743** (2016) 012002 doi:10.1088/1742-6596/743/1/012002

its industrial application. In addition, Lu *et al.* found experimentally that Zn(Co,Ga)O films are ferromagnetic. The authors also stated that the Zn(Co,Ga)O films exhibit enhanced n-type conduction in comparison to that of ZnO:Co. Furthermore, they found that the sp-d exchange interactions between O, Zn and Co atoms produce optical band gap variations in the Zn(Co,Ga)O films when compared to that of Co-doped ZnO [11]. However, no theoretical studies of the electronic structure and magnetic properties of (Ga,Co) co-doped ZnO system have been conducted in the crystallization wurtzite phase. Therefore, in this paper we have studied the electronic structure and ferromagnetic orderings in $Ga_xCo_{0.056}Zn_{1-(x+0.056)}O$ (x = 0.029, 0.059) using first principles calculations.

2. Computational methods

The calculations were performed using the first principles pseudo-potential method in the framework of the spin density-functional theory. Exchange and correlation effects were treated with generalized gradient approximation (GGA) implemented in the Perdew-Burke-Ernzerhof functional (PBE) [12]. The core electrons were described by the projector augmented wave (PAW) method [13, 14] wherein the d states for Ga, Co and Zn were included as valence electrons. The calculations were performed using vienna ab-initio simulation package (VASP) [15, 16]. The electron wave function was expanded in plane waves up to a cutoff energy of 500eV. A gamma-centered grid of 8×8×8 (3×3×4) k -point has been used to sample the irreducible Brillouin zone in the Monkhorst-Pack special scheme for the wurtzite structure with two Zn and two O atoms per primitive cell (3ax3ax2c supercell) [17]. Methfessel-Paxton smearing technique with a smearing width of 0.10eV was adopted [18]. These parameters ensure a convergence better than 1meV for the total energy. In order to investigate the magnetism of Gax $Co_{0.056}Zn_{1-(x+0.056)}O$ (x = 0.029, 0.059), we first studied the electronic and magnetic properties of $Co_{0.056}$ $Zn_{0.944}$ O using a 72-atom 3ax3ax2c supercell, based on a conventional ZnO wurtzite unit cell with the common lattice parameters a and c. The magnetic coupling between the doped Co atoms and the stable ground state of $Co_{0.056}$ $Zn_{0.944}$ O were studied by performing the total energy calculations corresponding to FM and AFM phases for both short and far Co-Co distances. The short and far Co-Co distances for $Co_{0.056}Zn_{0.944}O$ are ~3.3Å and 7.8Å, respectively, as shown in Table 1. Then, one and two Zn atoms were replaced in the most stable configuration (short Co-Co distances) of $Co_{0.056}Zn_{0.944}O$ supercell with one and two Ga ions (Figure 1) to represent concentrations of x = 0.029 and 0.059, respectively, in Co_x $Cu_{0.042}Zn_{1-(x+0.042)}O$. Next, the total energy of the $Ga_xCo_{0.056}Zn_{1-(x+0.056)}O$ (x = 0.029, 0.059) supercell was calculated with the two spin-polarized for Co and Ga atoms coupled in FM and AFM states. In all cases, the atomic positions were optimized.

Table 1. Calculated bond lengths between the two Co impurities (dC o–C o (Å)) before (in) and after (out) atomics relaxation, total magnetization (TM in μB/atom), total energy difference ($\Delta E=E_{AFM}-E_{FM}$ in meV), formation energies (E_f in eV) and ground state (GS) in ferromagnetic (FM) and antiferromagnetic (AFM) configurations for $Co_{0.056}Zn_{0.944}O$, $Ga_{0.029}Co_{0.056}Zn_{0.915}O$ and $Ga_{0.059}Co_{0.056}Zn_{0.885}O$ with GGA formalism.

$Ga_xCo_{0.056}Zn_{1-(0.056+x)}O$						
	$d_{Co\text{-}Co}$ (Å) In	out	TM ($\mu_B/atom$)	ΔE (meV)	E_f(eV)	GS
$Co_{0.056}Zn_{0.944}OGa_{0.029}$	3.26	3.21	3.09	17.10	3.34	FM
	7.77	7.77	0.00	-0.40	3.39	AFM
$Co_{0.056}Zn_{0.915}O$	3.26	3.20	2.98	12.70	2.62	FM
$Ga_{0.059}Co_{0.056}Zn_{0.885}O$	3.26	3.19	2.95	0.40	2.66	FM

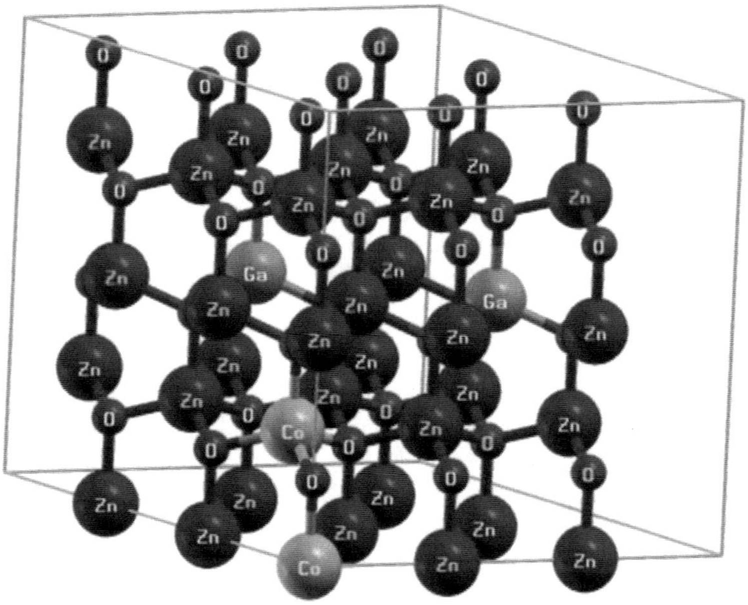

Figure 1. (Color online) 3a×3a×2c wurtzite supercell of $Ga_xCo_{0.056}Zn_{1-(x+0.056)}O$ for the Co- Co closest configuration. The gray, blue, green and red spheres represent respectively the Co, Zn, Ga and O atoms. Two Co and two Ga atoms are set at Zn lattice sites for $Ga_{0.059}Co_{0.056}Zn_{0.885}O$ supercell.

3. Results and discussion

3.1. Structural Properties of ZnO semiconductor and Magnetic properties of $Co_{0.056}Zn_{0.944}O$ with GGA formalism

At ambient conditions, ZnO crystallizes in the hexagonal close-packed wurtzite structure. In this primitive cell there are two units of ZnO, tetrahedrally coordinated each, where four atoms of oxygen surround each atom of zinc. The experimental lattice parameters a, c/a and u of wurtzite ZnO are, respectively, a = 3.253Å, c/a = 1.603Å and u = 0.382c [19]. Our PBE results overestimate both a and c quantities by 1.107% and 1.497%, respectively, and underestimate the u parameter by 0.523%. Table 1 displays that the ground state (GS) of $Co_{0.056}Zn_{0.944}O$ system.

Is ferromagnetic (FM) for the nearest Co-Co distance (~ 3.3Å) with a net magnetic moment and FM energy difference ($\Delta E=E_{AFM}-E_{FM}$ in meV) of ~3.00μB per Co atom and 17.10meV, respectively. On the contrary, for the farthest Co-Co distance (~7.8Å) the ground state of $Co_{0.056}Zn_{0.944}O$ is antiferromagnetic (AFM) with $\Delta E = -0.4$meV. The formation energy (E_f) for the FM and AFM ground state for $Co_{0.056}Zn_{0.944}O$ are, respectively, ~3.3 and ~3.4eV. Previous theoretical and experimental studies reported of Co-doped ZnO have been diverging. Some studies report Curie temperatures while others paramagnetism, superparamagnetism or anti-ferromagnetism [9]. In addition Sik *et al.* stated that the realization of the FM long range order in Co-doped ZnO films is controversial [10]. Therefore, codoping Co-doped ZnO with trivalent Ga dopants could be efficient to stabilize the ferromagnetic order in Co-doped ZnO.

3.2. Magnetic properties of $Ga_xCo_{0.056}Zn_{1-(x+0.056)}O$ (x = 0.029, 0.059) with GGA scheme

In order to study the magnetic properties of $Co_{0.056}Zn_{0.944}O$ system with GGA for the short Co-Co distances (~3.3Å), which was the most stable configuration for Co-doped ZnO, we first replace one (two) Zn atom in the $Co_{0.056}Zn_{0.944}O$ host by one (two) Ga atom to represent a x concentration of 0.029

(0.059). Then, the total energy of the $Ga_xCo_{0.056}Zn_{1-(x+0.056)}O$ (x = 0.029, 0.059) supercell was calculated with the two spin-polarized for Co and Ga atoms coupled in FM and AFM states. Table 1 displays that the FM ground state is the most stable configuration for $Ga_xCo_{0.056}Zn_{1-(x+0.056)}O$ (x = 0.029) system with ΔE =~ 13meV. However, for $Ga_xCo_{0.056}Zn_{1-(x+0.056)}O$ (x = 0.059) system, Table 1 shows that the FM ground state almost vanishes (ΔE =~ 0.40meV). From Table 1 we can also see that the formation energy for $Ga_xCo_{0.056}Zn_{1-(x+0.056)}O$ (x = 0.029, 0.059) is lower than that of $Co_{0.056}Zn_{0.944}O$. This means that (Ga,Co) co-doped ZnO is more energetically stable than $Co_{0.056}Zn_{0.944}O$ system. On the other hand, the total magnetic moment per Co atom has almost the same value (~ 3.00μB) per supercell for both $Ga_{0.029}Co_{0.056}Zn_{0.915}O$ and $Co_{0.056}Zn_{0.944}O$ systems.

3.3. TDOS and PDOS of $Co_{0.056}Zn_{0.944}O$ and $Ga_xCo_{0.056}Zn_{1-(x+0.056)}O$ (x = 0.029,0.059) with GGA scheme

In order to provide fundamental insight into the interaction of both Co with ZnO and Ga with Co-doped ZnO, and how these interactions can produce an induced magnetism, we calculated the total density of states (TDOS) for $Co_{0.056}Zn_{0.944}O$ and $Ga_xCo_{0.056}Zn_{1-(x+0.056)}O$ (x = 0.029,0.059), and partial density of states (PDOS) for 3d-Co, 2p-O, 3d-Ga, 2s-Ga and 3d-Zn states in the FM configuration, as shown in Figures 2(a), 2(b) and 2(c), respectively. When Co replaces Zn in ZnO, the nearly tetrahedral crystal field formed by O ions splits the five fold degenerate 3d-Co states into the twofold ($d_{x^2-y^2}$, d_{z^2}) and threefold (d_{xy}, d_{xz}, d_{yz}) degenerated e and t_2 states, where the e states are lower in energy than the t_2 states. TDOS of $Co_{0.056}Zn_{0.944}O$ (Figure 2(a)) displays some states in the minority band which lie both below and near the Fermi level and on the Fermi level. Thus, the system is nearly halfmetallic. Similarly, PDOSs for 3d-Co and 2p-O of $Co_{0.056}Zn_{0.944}O$ show that some states in the minority band lie at the same energy levels. Above the Fermi level, in the minority band of TDOS, between ~ 0.8 and 1.5eV, some peaks are shown. These peaks are also shown in PDOS for 3d-Co and 2p-O at the same energy levels. The fact that 3d-Co and 2p-O states are located at the same energy level and look alike suggests a p-d hybridization between 3d-Co and 2p-O states. The states shown in the minority band of PDOS for 3d-Co which lie below (above) and on the Fermi level correspond to e (t_2) states. The up-spin Co d states are fully occupied, while the down-spin d states are partially occupied. As we can see in TDOS of $Co_{0.056}Zn_{0.944}O$ (Figure 2(a)) the empty minority states above Fermi level correspond to t_2 states. On the other hand, the valence electron configuration of Co is $3d^74s^2$. If Zn^{2+} is replaced by Co^{2+}, the configuration of Co is $3d^7$. Therefore, as we added 2Co, we expected a total magnetic moment of no more than 6.0 μB per supercell. Indeed, we obtained a total magnetic moment of ~6.0 μB per supercell (see Table 1), near to the predicted value. Similar results were found by Spaldin. The author reports by a DFT study of $Co_xZn_{1-x}O$ (x = 0.0625 and 0.125) [8] a total magnetization of ~6.0 μB per unit cell. Park *et al.* [10] reported a total magnetic moment per Co atom of 3.25 μB for $Co_xZn_{1-x}O$ (x = 0.0625) by using LSDA approach. Figures 2(a) and 2(b) display the calculated TDOS and PDOS for 3d-Co, 2p-O, 3d-Ga, 2s-Ga and 3d-Zn states of $Ga_xCo_{0.056}Zn_{1-(x+0.056)}O$ with x = 0.029 and 0.059, respectively, in the FM configuration. As we can see in Figures 2(a) and 2(b), when Ga replaces Zn in $Co_{0.056}Zn_{0.944}O$, the $Co_{0.056}Zn_{0.944}O$ band gap is modified due to the Ga(4s)-O(2p)-Co(3d) exchange interactions. We can see in the PDOS of 4s-Ga for $Ga_xCo_{0.056}Zn_{1-(x+0.056)}O$ (x = 0.029, 0.059) that some states in both majority and minority bands lie on the Fermi level. As a result, the nearly halfmetallic nature of $Co_{0.056}Zn_{0.944}O$ changes to metallic due to 4s-Ga partial occupation band. On the other hand, the valence electron configurations of Ga is $3d^{10}4s^24p^1$. If Zn^{2+} is replaced by Ga^{2+}, the configuration of Ga is $3d^{10}4s^1$. Through Bader analysis [20, 21, 22], we found a total valence charge of 11.3e and no magnetization on the Ga atom for $Ga_xCo_{0.056}Zn_{1-(x+0.056)}O$ (x = 0.029, 0.059). In comparison to the ideal values of 13e for Ga, the Ga adopts approximately ionization states of 2^+. As $3d^{10}$-Ga orbitals are all fully occupied, the $4s^1$-Ga enhances the n-type conduction of $Ga_xCo_{0.056}Zn_{1-(x+0.056)}O$ system in comparison to that of Co-doped ZnO. These results agree with those found experimentally by Lu *et al.* The authors reported that the Zn(Co,Ga)O films exhibit enhanced n-type conduction in comparison to that of Co-doped ZnO.

Furthermore, they found that the sp-d exchange interactions between O, Ga and Co atoms produce optical band gap variations in the Zn(Co,Ga)O films when compared to that of Co-doped ZnO [11].

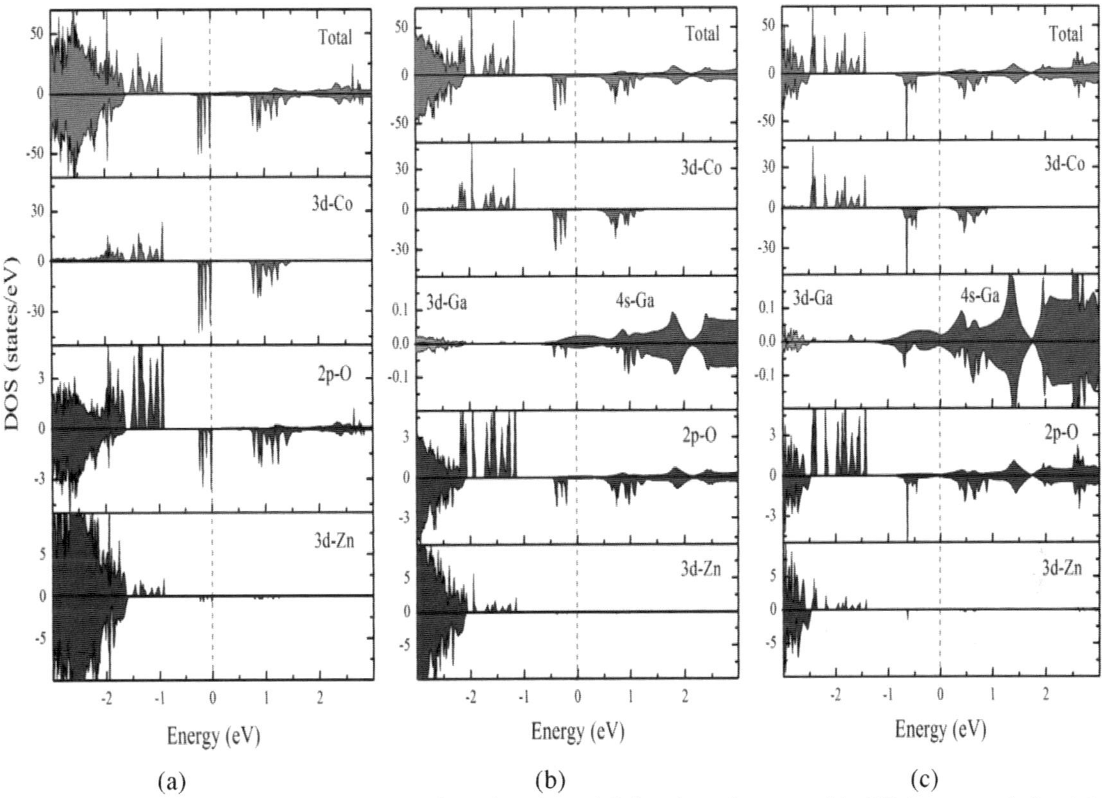

Figure 2. (Colour online) Comparison of total and partial density of states with GGA approach for (a) $Co_{0.056}Zn_{0.944}O$ (b) $Ga_{0.029}Co_{0.056}Zn_{0.915}O$ and (c) $Ga_{0.059}Co_{0.056}Zn_{0.885}O$ in ferromagnetic configuration.

4. Conclusions

The electronic structure and magnetic properties of $Ga_xCo_{0.056}Zn_{1-(x+0.056)}O$ (x = 0,0.029 and 0.059) system have been studied using first-principles calculations. A ferromagnetic ground state with total magnetic moment of ~3.0 μB per Co atom has been found for the closest Co-Co distance in $Co_{0.056}Zn_{0.944}O$ system. On the contrary, an antiferromagnetic ground state was found for the farthest Co-Co distance. Formation energy calculations display that $Ga_xCo_{0.056}Zn_{1-(x+0.056)}O$ (x = 0.029,0.059) system is more energetically stable than $Co_{0.056}Zn_{0.944}O$. It has also been found that Ga-doping reduces the $Co_{0.056}Zn_{0.944}O$ band gap due to the Ga(4s)-O(2p)-Co(3d) exchange interactions, which is in good agreement with the experimental data. Through Bader analysis, we found that the $4s^1$-Ga orbital enhances the n-type conduction of $Ga_xCo_{0.056}Zn_{1-(x+0.056)}$ O system in comparison to that of Co-doped ZnO. Results also show that $Ga_{0.029}Co_{0.056}Zn_{0.915}O$ is still ferromagnetic with a total magnetic moment of ~3.0 μB per Co atom. Conversely, ferromagnetic ground state in (Ga,Co) co-doped ZnO vanishes as Ga concentration increases.

Acknowledgements

This work has been carried out with the financial support of Universidad del Norte and Colciencias (Administrative Department of Science, Technology and Research of Colombia) under "Convocatoria 658 - Convocatoria para proyectos de investigaci´on en ciencias b´asicas año 2014".

References

[1] Huang H, Fang G, Mo X, Long H, Yuan L, Dong B, Meng X and Zhao X 2009 *IEEE Electron Device Lett.* **30** 1063-1065
[2] Gonzalez R, Lopez-Perez W and Rodriguez J 2009 *J. Magn. Magn. Mater.* **321** 2547-2549
[3] Pearton S, Norton D, Heo Y, Tien L, Ivill M, Li Y, Kang B, Ren F, Kelly J and Hebard A 2006 *J. Electron. Mater.* **35** 862-868
[4] Pandey B, Ghosha S, Srivastava P, Avasthi D, Kabiraj D and Pivin J 2008 *J. Magn. Magn. Mater.* **320** 3347-3351
[5] Zutic I, Fabian J and Sarma S 2004 *Rev. Mod. Phys.* **76** 323-410
[6] González A, López W, Barragán D and González R 2015 *J. Supercond. Nov. Magn.* **28** 318592
[7] Liang W, Tingjun H, Yi W, Yanfei Z, Zhenyu G, Youyong L and Shuit-Tong L 2012 *J. Alloys Compd.* **541** 2505
[8] Spaldin N 2004 *Phys. Rev. B* **69** 125201
[9] Jun Z *et al.* 2002 *J. Cryst. Growth* **548** 237-239
[10] Park S and Min B 2003 *Phys. Rev. B* **68** 224436
[11] Lu B, Wanga Y, Li W, Zhang, Ye Y, Zhang L and Ye Z 2015 *J. Magn. Magn. Mater.* **374** 278-282
[12] Perdew J, Burke K and Ernzerhof M 1996 *Phys. Rev. Lett.* **77** 3865-3868
[13] Blochl P 1994 *Phys. Rev. B* **50** 17953-17979
[14] Kresse G and Joubert D 1999 *Phys. Rev. B* **59** 1758-1775
[15] Kresse G and Furthmu¨ller J 1996 *Comput. Mat. Sci.* **6** 15-50
[16] Kresse G and Furthmu¨ller J 1996 *Phys. Rev. B* **54** 11169-11186
[17] Monkhorst H and Pack J 1976 *Phys. Rev. B* **13** 5188-5192
[18] Methfessel M and Paxon A 1989 *Phys. Rev. B* **40** 3616-3621
[19] Schulz H and Thiemann K 1979 *Solid State Commun.* **32** 783-785
[20] Henkelman G, Arnaldsson A and Jonsson H 2006 *Comput. Mater. Sci.* **36** 354-360
[21] Sanville E, Kenny S, Smith R and Henkelman G 2007 *Comput. Chem.* **28** 899-908
[22] Tang W, Sanville E and Henkelman G 2009 *Phys.: Condens. Matter.* **21** 084204-084210

Journal of Physics: Conference Series **743** (2016) 012003 doi:10.1088/1742-6596/743/1/012003

Relative phase and physical properties of CrN/AlN multilayer: A DFT study

E F Cudris[1], J H Díaz F[1] and M J Espita R[1]
[1] Universidad Distrital Francisco José de Caldas, Bogotá, Colombia.

E-mail: ecudronobel@gmail.com

Abstract. Using first principles total-energy calculations within the framework of density functional theory, we studied the relative stability and the structural and electronic properties of multilayer CrN/AlN in the sodium chloride (NaCl), cesium chloride (CsCl), nickel arsenide (NiAs), zinc-blende, and wurtzite structures. The calculations were carried out using the method based on pseudopotentials, employed exactly as implemented in Quantum-ESPRESSO code. Based on total energy minimization, we found that the minimum global energy of CrN/AlN is obtained for the zincblende structure. Additionally, at high pressure our calculations show the possibility of a phase transition from the zincblende to NaCl structure. For the zincblende phase, the density of states analysis reveals that the multilayer exhibits a half-metallic behavior with a magnetic moment of $3.0\mu_{\beta}$/Cr-atom. These properties come essentially from the polarization of Cr-d and N-p states that cross the Fermi level. Due to these properties, the multilayer can potentially be used in the field of spintronics or spin injectors.

1. Introduction

The group III nitrides, such as aluminum nitride (AlN), have been the subject of many intense investigations during the last few years. This interest is partially motivated by their promising applications in the field of optical devices in the violet region [1]. They are also used as electronic packaging material and are applied to optical disks as well as lithographic photo masks [2-4]. AlN stabilizes in the wurtzite structure in bulk form, but can also be grown in the zincblende structure under appropriate conditions [5, 6]. On the other hand, chromium nitride stabilizes in the zincblende structure [7]. This material has excellent properties, such as a high degree of hardness, brittleness, a high melting temperature, good electrical-thermal conductivity and superconductivity, which makes it suitable for many important technological applications, for example hard coatings, cutting tools, and punching machines [8-13]. Additionally, over the last few years there has been a great deal of interest in developing new and improved ultrahard materials using the group III nitrides and a metal transition ion. Experimental studies show that these superhard compounds can be grown; for example, a TiN/GaN multilayer was grown via reactive pulsed laser deposition [14, 15], while an AlN/VN multilayer was grown via reactive sputtering [16]. Additionally, the combination of the semiconductor properties of AlN and the magnetic properties of the transition metal ions is currently of great interest because of its potential applications in diluted magnetic semiconductors (DMS) [17, 18]. For this reason, in this paper we report a detailed theoretical calculation of the structural and electronic properties of CrN/AlN in order to determine the energetically most favorable structure. We present the results of the calculated structural and magnetic properties in the zincblende, wurtzite, rock-salt, CsCl,

Content from this work may be used under the terms of the Creative Commons Attribution 3.0 licence. Any further distribution of this work must maintain attribution to the author(s) and the title of the work, journal citation and DOI.
Published under licence by IOP Publishing Ltd

and NiAs structures, and finally, we present a study of the electronic properties of the most favorable phase, the zincblende structure.

2. Computational method

The calculations were performed within the DFT framework using the Quantum Espresso package [19]. The correlation and exchange effects of the electrons were dealt with using the generalized gradient approximation (GGA) of Perdew, Burke, and Ernzerhof (PBE) [20]. Electron-ion interactions were studied using the pseudopotential method [21, 22]. The electron wave functions were expanded into plane waves with a kinetic-energy cutoff of 40Ry. For the charge density, a kinetic energy cutoff of 400Ry was used. The calculations were performed taking into account the spin polarization due to the presence of the Cr atom. To calculate the lattice constant, the bulk modulus, and the total energy of each studied superlattice, the calculated data were fit to the Murnaghan equation of state [23]. The CrN/AlN multilayer was modeled according to the special quasirandom structures approach [24], and the disorder aspects were ignored.

To determine the structural properties in the ground state, such as the lattice constant (a_0), bulk modulus (B_0) and total energy (E_0) of the CrN/AlN multilayer in the CsCl, NaCl, NiAs, zincblende, and wurtzite structures, the total energy was calculated as a function of the volume and the results were fit to the Murnaghan equation of state, Equation (1).

$$E(V) = E_0 + \frac{B_0 V}{B_0'}\left[\frac{\left(V_0/V\right)^{B'}}{B_0'-1}+1\right] - \frac{B_0 V_0}{B_0'-1} \qquad (1)$$

where B_0 is the bulk modulus, and its first derivative is B'_0.

3. Results and discussions

3.1. Structural properties

Figure 1 shows the energy-volume curves for the CrN/AlN multilayer. It can be noted that the five structures considered in this study are metastable, because for each of the energy-volume curves there is a minimum energy value. In Figure 1, we can clearly see that the zincblende phase is the energetically most favorable structure, because it has the lowest minimum energy value. Additionally, we can see that the curve corresponding to the zincblende phase crosses the curve of the NaCl structure of smallest equilibrium volume. This indicates a high-pressure phase transition from the zincblende to the NaCl structure. In order to analyze the phase transformation under high pressure, the Gibbs free energy equation $G=E+PV-TS$ can be used. Since the calculations are for the ground state (T=0°K), the last term of the Gibbs energy equation can be neglected, and by working with the enthalpy, the Gibbs free energy equation reduces to $H=E+PV$ [25, 26]. We use this equation for the five crystal structures considered in this paper; however, we cannot exclude the possible existence of other stable or metastable structures for the CrN/AlN multilayer.

Figure 2 shows the enthalpy as a function of pressure for the zincblende and NaCl phases. We note that the compound will transform from a zincblende structure to NaCl at a transition pressure of P_T=2.5GPa. Note that before the transition ($0<P<P_T$), the lowest values of enthalpy correspond to the zincblende phase. This shows the relative stability of the zincblende phase within this pressure range, while for higher values of the transition pressure ($P>P_T$), the NaCl phase has the lowest values of enthalpy. Therefore, within this pressure range, NaCl is the most stable phase. At the transition pressure, the enthalpies of both structures are equal. Across the transition, there is a volume reduction of about 17.75%, from 42.899Å^3 to 35.284Å^3. A similar transition phase was reported by Casiano *et al.* in their study of the relative stability of the CrN compound [27].

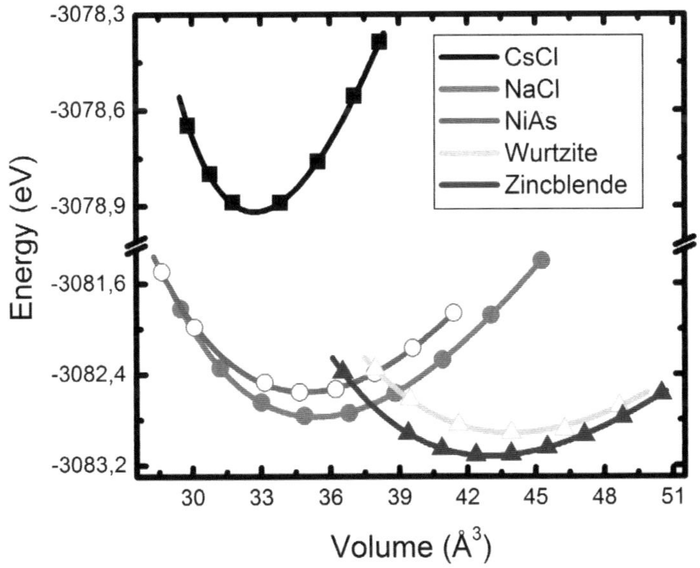

Figure 1. Total energy as a function of the volume of the CrN/AlN multilayer in the FM states.

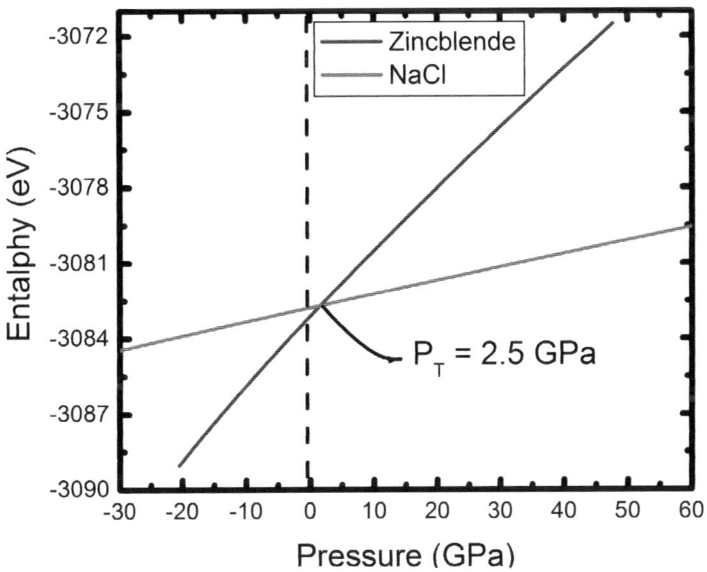

Figure 2. Enthalpy as a function of pressure in the NaCl and zincblende phases of CrN/AlN multilayer.

Table 1 shows the structural parameters of CrN/AlN, calculated for the phases CsCl, NaCl, NiAs, wurtzite, and zincblende.

Table 1. Structural parameters lattice constant (a_0), bulk modulus (B_0) and total energy (E_0) of the superlattices.

Phase	a_0 (Å)	c/a	V_0 (Å^3)	B_0 (GPa)	E_0 (eV)	μ (μ_β)
CsCl	2.5401	-	32.733	266.082	-3078.917	~2.8
NaCl	2.9240	1.414	35.284	188.988	-3082.645	~3.0
NiAs	4.4146	-	42.899	177.491	-3082.554	~3.0
Zincblende	2.8753	1.696	34.924	230.821	-3083.117	~4.1
Wurtzite	3.1561	1.613	43.927	172.011	-3082.774	~3.0

We can see that the phase with the lowest value of energy corresponds to the zincblende structure. Additionally, we can see that the value of the bulk module (230.821GPa) of the CrN/AlN multilayer in the zincblende phase is higher than the bulk modules of CrN (204.5GaP) [27] and AlN (191.85GPa) [28], which confirms that they are quite rigid, making them good candidates for possible applications in devices operated at high temperature and high power, as well as in hard coatings.

3.2. Electronic properties

Figure 3(a) shows the total and partial spin-polarized density of states of the CrN/AlN multilayer in the zincblende phase. Figure 3 shows that they are half-metallic and ferromagnetic. This result occurs because in the valence band close to the Fermi level, the majority spins (spin-up) are metallic, and the minority spins (spin-down) are semiconductors. These compounds have a spin polarization of 100% of the conduction carriers in the ground state, which is a requirement for spin injectors [29, 30].

This finding suggests that this compound can be efficiently used for spin injectors. Figure 3(a) shows that in the valence band near the Fermi level, the spin-up density of the majority spins (up) is mainly dominated by the Cr-3d states, and to a lesser extent by the N-2p states, which cross the Fermi level. The magnetic properties essentially come from the chromium atom, and can be attributed to the Cr^{3+} configuration, because the Cr atom in the superlattice donates three electrons, and the Cr atom retains two valence electrons (the electronic configurations Cr^{3+}=:[Ar]$3d^3$). These valence electrons couple ferromagnetically, and as a result the three electrons produce a total magnetic moment of $3\mu_\beta$/atom-Cr. Figure 3(b) shows the electronic band structure for up and down spin. The band structure confirms that the multilayer has a half-metallic character, because the minority spin down (dn) states preserve a band gap, but in the majority spin (up) states there is a penetration toward the prohibited energy zone of the 3d-Cr states in a greater proportion and the 2p-N and 2p-Al states in a lesser proportion.

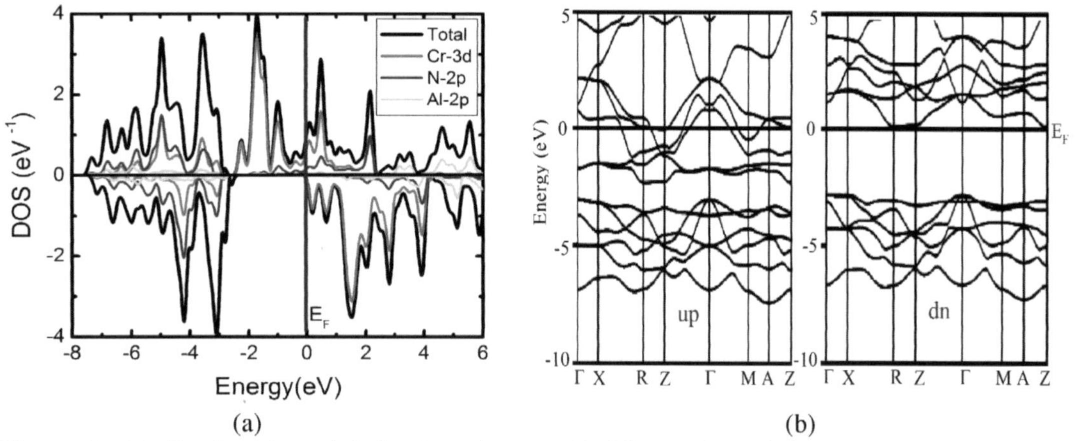

Figure 3. (a) Total and partial density of state (b) Electronic band energy for the CrN/AlN multilayer.

4. Conclusions

We reported first-principles calculations to determine the structural, electronic, and magnetic properties of a AlN/GaN multilayer for the CsCl, NaCl, NiAs, zinc-blende, and wurtzite structures, within the framework of density functional theory using the approach of the generalized gradient approximation (GGA). From the computation of the total energy versus unit-cell volume, the zincblende structure was found to be favored. We found that the transition pressure for the zincblende to the NaCl phase was about 2.5GPa. The calculated values of the bulk modules are quite high; therefore, the multilayers are quite rigid, which makes them attractive for potential applications at high temperatures and for hard coatings. On the basis of the density of states for the zincblende phase, we found that the multilayer exhibits a half-metallic behavior with a magnetic moment of $3\mu_{\beta}$/Cr-atom. The ground-state ferromagnetic behavior comes from the polarization of the C-3d and N-2p orbitals, which cross the Femi level. These compounds are good candidates for potential applications in spintronics and can be used as spin injectors.

References

[1] Nakamura S 1997 *Solid State Communications* **102** 237
[2] Wagner JM and Bechstedt F 2002 *Physical Review B* **66** 115202
[3] Nakamura S, Senoh M, Iwasa N and Nagahama S 1995 *Applied Physics Letters* **67** 1868
[4] Binari SC, Kruppa W, Dietrich HB, Kelner G, Wickenden AE and Freitas JA 1997 *Solid State Electron* **41** 1549
[5] Christensen NE and Gorczyca I *1994 Phys. Rev. B* **50** 4396
[6] Bi WG and Tu CW 1996 *J. App. Phys.* **80** 1934
[7] Zhou L, Holec D and Mayrhofer PH 2013 *J. Phys. D: Appl. Phys.* **46(36)** 365301.
[8] Jhi S-H, Ihm J, Louie SG and Cohen ML 1999 *Nature* **399** 132
[9] McMillan PF 2002 *Nat. Mater.* **1** 19.
[10] Maruyama T and Morishita T 1996 *Appl. Phys. Lett.* **69** 890
[11] Yamanaka S, Hotehama K and Kawaji H 1998 *Nature* **92** 580
[12] Zerr A, Miehe G and Boehler R 2003 *Nat. Mater.* **2** 185
[13] Chhowalla M and Unalan HE 2005 *Nat. Mater.* **4** 317
[14] Rawat V and Sands T 2006 *Journal of Applied Physics* **100** 064901
[15] Espitia MJ, Ortega L and Casiano J 2011 *Revista Colombiana de Física* **43** 408
[16] Kim IW, Li Q, Marks LD and Barnett SA 2001 Appl. Phys. Lett. **78** 892
[17] Yao G, Fan G, Zheng S, Chen J, Zhou D, Li S, Zhang Y and Su S 2012 *Optical Materials* **34** 593
[18] Sato K, Dederics PH and Yoshida K 2003 *Europhysics Letters* **61** 40300191
[19] Giannozzi P, Baroni S and Bonin N 2009 *J. Phys. Condens. Matter* **21** 395502 http://www.quantum-espresso.org
[20] Perdew J, Burke K and Ernzerhof M 1996 *Physical Review Letter* **77** 3865
[21] Vanderbilt D 1990 *Phys. Rev. B* **41** 7892
[22] Laasonen K, Pasquarello A, Car R, Lee C and Vanderbilt D 1993 *Phys. Rev. B* **47** 10142
[23] Murnaghan FD 1944 *Proc Natl Acad Sci U S A* **30** 244
[24] Zunger A, Wei SH, Ferreira L and Bernard JE 1990 *Physical Review Letters* **65** 353
[25] Espitia-Rico MJ, Díaz-Forero JH and Castillo-Méndez LE 2015 *Revista Facultad de Ingeniería de la Universidad de Antioquía* **76** 143
[26] Báez Cruz RE, Ortega López C and Espitia MJ 2013 *Revista Ingeniería y Ciencia* **17** 163
[27] Casiano G, López-Pérez W, Ortega C and Rodríguez JA 2009 *Revista Colombiana de Física* **42** 580
[28] Beloufa A, Bensaad Z, Soudini B-A, Sekkal N, Bensaad A, Abid H 2009 *International Journal Nanoelectronics and Matrials* **2** 11
[29] Heddar H, Zaoui A and Ferhat M 2013 *Superlattices and Microstructures* **53** 16
[30] Vargas-Hernández C, Espitia M, Báez R E 2015 *Computational Condensed Matter* **4** 1

Journal of Physics: Conference Series **743** (2016) 012004 doi:10.1088/1742-6596/743/1/012004

Research of the structural and electronic properties of VN/AlN/VN and AlN/VN/AlN based on DFT calculation

J H Díaz F[1], M J Espitia R[1] and J A Rodríguez Martínez[2]
[1] Universidad Distrital Francisco José de Caldas, Bogotá, Colombia.
[2] Universidad Nacional de Colombia, Bogotá, Colombia.

E-mail: jhdiazf@udistrital.edu.co

Abstract. We investigated the structural, electronic, and magnetic properties of AlN/VN/AlN and VN/AlN/VN interlayers in the wurtzite structure, via first-principles calculations using the psuedopotential method within density functional theory (DFT) as implemented in the Quantum ESPRESSO code. The total energy calculation showed that the VN/AlN/VN interlayer is energetically more favorable than the AlN/VN/AlN one. Analysis of the density of states showed that the interlayers exhibit a metallic behavior that essentially comes from the hybridization and polarization of states V-d and N-p across the Fermi level. The interlayers exhibit magnetic properties, with a magnetic moment of $4\mu_\beta$/cell. Due to these properties, the superlattices can potentially be used in the field of spintronics.

1. Introduction

Aluminum nitride (AlN) normally crystallizes in the wurtzite structure [1, 2]. This material has recently attracted the attention of researchers due to its excellent properties, and has been widely used in the design of light-emitting diodes (LEDs) and laser diodes (LDs) [3] and in electronic packaging material, and has been applied to optical disks as well as lithographic photo masks [4-6]. Due to its stability at high temperature, high degree of thermal conductivity, low thermal expansion, and high resistance to gases and chemicals [7], it has been used in many electronic devices that must function at high temperature and high power and in corrosive environments. Over the last few years, AlN has received extensive attention because of its possible use as a diluted magnetic semiconductor (DMS) with potential applications in the field of spintronics. There is also interest in various applications in spintronics (e.g. electro-optical switches and spin injectors) [8]. In the present paper, we study the structural and electronic properties of AlN/VN/AlN and VN/AlN/VN interlayers in the wurtzite structure.

2. Computational method

Calculations were performed using periodic density functional theory (DFT) as implemented in the plane-wave self-consistent field (PWscf) code in the Quantum ESPRESSO package [9]. The exchange and correlation energies were modeled according to the generalized gradient approximation (GGA) with the Perdew Burke Ernzerhof (PBE) gradient-corrected functional [10]. Electron-ion interactions were dealt with using the pseudopotential method [11, 12]. The electron wave functions were expanded into plane waves with a kinetic-energy cutoff of 40Ry. For the charge density, a kinetic energy cutoff of 400Ry was used. A 6×6×3 Monkhorst-Pack mesh [13] was used to generate the k-

Content from this work may be used under the terms of the Creative Commons Attribution 3.0 licence. Any further distribution of this work must maintain attribution to the author(s) and the title of the work, journal citation and DOI.
Published under licence by IOP Publishing Ltd

Journal of Physics: Conference Series **743** (2016) 012004 doi:10.1088/1742-6596/743/1/012004

points in the supercell. To calculate the lattice constant, the bulk modulus, and the total energy of each studied compound, the calculated data were fit to the Murnaghan equation of state [14], Equation (1).

$$E(V) = E_0 + \frac{B_0 V}{B_0^{'}}\left[\frac{\left(V_0/V\right)^{B'}}{B_0^{'}-1}+1\right]-\frac{B_0 V_0}{B_0^{'}-1} \qquad (1)$$

where B_0 is the bulk modulus and its first derivative is $B^{'}_0$, V_0 is the equilibrium volume of the cell, and E_0 is the binding energy. In other to study the relative stability of AlN/VN/AlN and VN/AlN/VN interlayers, the energy of formation was calculated. For a ternary compound, the formation energy is defined as the difference between the total energy of the ternary M = AlN/VN/AlN (or M=VN/AlN/VN) and the total energy of the binary compounds VN-NaCl and AlN-wurtzite in their ground state. Therefore, the formation energy is given by Equation (2) [15, 16].

$$\Delta E_f = E_M^{wurtzita} - (1-x)\, E_{VN}^{NaCl} - x E_{AlN}^{wurtzita} \qquad (2)$$

The interlayer is modeled as sandwiching two layers of AlN between two layers of VN (or VN between two layers of AlN), and was modeled according to the special quasirandom structures approach [17]. The disorder aspects were ignored.

3. Results and discussions

3.1. Structural properties

In order to obtain the structural properties in the ground state, such as the equilibrium lattice constant (a_0), the bulk modulus (B_0), the c/a ratio, the total energy (E_0), and the magnetic moment (μ) per cell of AlN/VN/AlN and VN/AlN/VN interlayers in the wurtzite structure, the total energy of the compound was minimized as a function of cell volume. The results were fit to the Murnaghan equation of state, Equation (1). Figure 1 shows the total energy-volume curves for the interlayers.

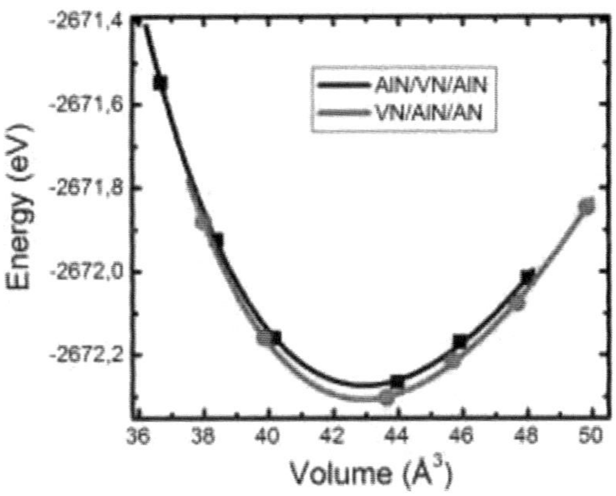

Figure 1. Total energy as a function of the volume for the AlN/VN/AlN and VN/AlN/VN interlayers.

The VN/AlN/VN interlayer has a lower minimum value of total energy than the AlN/VN/AlN interlayer; therefore, the VN/AlN/VN interlayer is energetically more stable.

The lattice constant, the c/a value, the bulk modulus (B_0), the total energy (E_0), and the magnetic moment (μ_β) per cell are listed in Table 1.

Table 1. Structural parameters: lattice constants (a_0) and (c), bulk modulus (B_0), total energy (E_0), and magnetic moment per cell.

Interlayer	a_0 (Å)	c (Å)	B_0 (GPa)	E_0 (eV)	μ (μ_β)/cell
AlN/VN/AlN	3.0874	3.370	177.532	-2672.2732	4.0
VN/AlN/VN	3.0869	3.371	181.310	-2672.2965	4.0

We note that the values of the bulk modulus of the interlayers are high. These values are close to the bulk module of AlN [17] and VN in the wurtzite structure [18]; hence the interlayers are quite rigid, making them good candidates for possible applications in devices operated at high temperature and high power, as well as in hard coatings. The energy difference between VN/AlN/VN and AlN/VN/AlN is -273.2meV, so in the ground state the VN/AlN/VN interlayer has lower energy than the AlN/VN/AlN. In order to verify the relative stability of the VN/AlN/VN and AlN/VN/AlN interlayers, we calculated the corresponding formation energy, using Equation (2). We calculated the total energy E_{VN}^{NaCl} and $E_{AlN}^{wurtzite}$ in their ground states. The values are -2631.48eV and –2715.234eV, respectively. Table 2 shows the calculated values of the formation energy E_f.

Table 2. Formation energy.

Compound	ΔE_f (eV)
VN/AlN/VN	1.060
AlN/VN/AlN	1.084

According to the results shown in Table 2, the smallest value of the energy of formation corresponds to the VN/AlN/VN interlayer; therefore, it is the most energetically stable structure. Additionally, the moderate formation-energy values indicate that the compounds can easily be grown experimentally. These results for the energy of formation are important, because if one knows these values, growing conditions can be improved, and therefore a VN/AlN/VN interlayer of excellent quality can be grown.

Figure 2 shows the crystal structure of the VN/AlN/VN interlayer obtained after structural relaxation. For both the VN/AlN/VN interlayer and the AlN/VN/AlN interlayer, the space group obtained is the same: tetragonal structure 156 (P3m1).

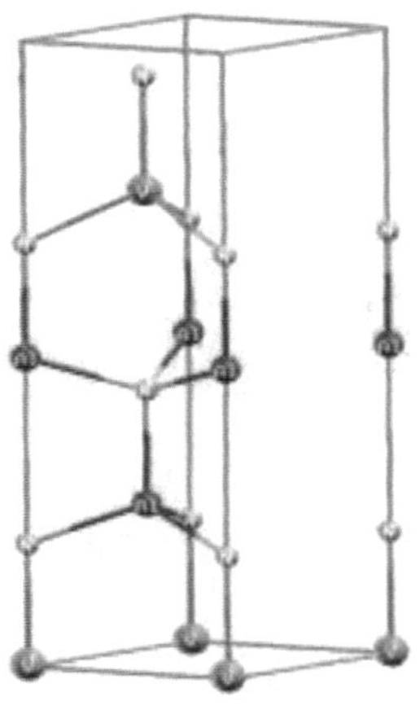

Figure 2. Unit cell of the VN/AlN/VN interlayer after structural relaxation.

Journal of Physics: Conference Series **743** (2016) 012004 doi:10.1088/1742-6596/743/1/012004

3.2. Electronic properties

Figure 3 shows the total density of states (TDOS) and partial density of states (PDOS) of the orbitals that contribute most near the Fermi level of the VN/AlN/VN interlayer. The calculations were performed with spin polarization up and down.

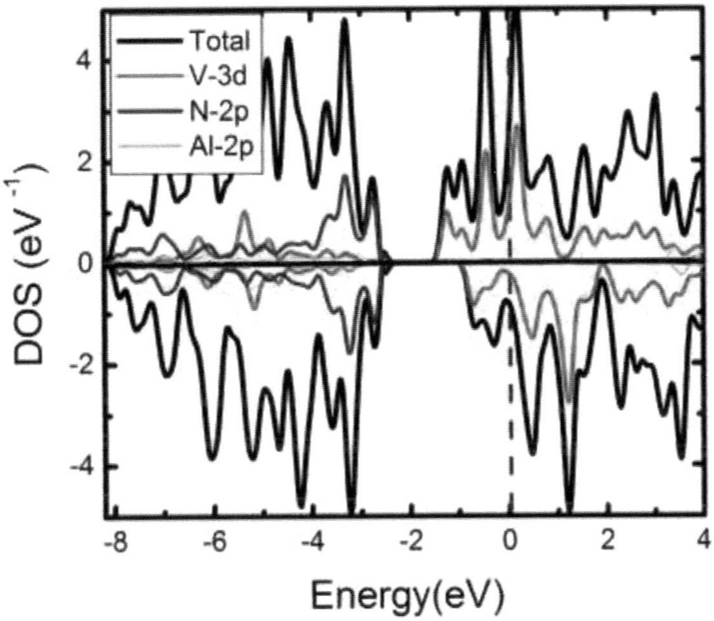

Figure 3. Total and partial density of states of the VN/AlN/VN interlayer in the wurtzite structure.

The interlayer exhibits a metallic behavior due to the fact that the valence orbitals cross the Fermi level. This metallic behavior is determined by the V-3d state. The interlayer exhibits magnetic properties, with a total magnetic moment of $4\mu_\beta$ per supercell. Since there are two vanadium atoms in the supercell, this corresponds to $2\mu_\beta$/V-atom. The magnetic moment is due to the V^{3+} configuration (with electronic configuration [Ar]$3d^2$). This implies that when the V atom is in the interlayer, it gives up three electrons, and two valence electrons remain (d^2 configuration). These valence electrons couple, and as a result the two electrons produce a total magnetic moment of $2\mu_\beta$/V atom. Touati *et al.* [19] reported a valence of V^{3+} for infrared luminescence in V-doped GaN samples grown with MOVPE on a sapphire substrate.

4. Conclusions

We carried out a study of AlN/VN/AlN and VN/AlN/VN interlayers using density functional theory (DFT) within the pseudopotential framework. The total energy calculation showed that the VN/AlN/VN interlayer is more energetically favorable than the AlN/VN/AlN one. The values of the bulk modules of the interlayers are high, close to the bulk modules of AlN and VN in the wurtzite structure. This confirms that they are quite rigid, making them good candidates for possible applications in devices operated at high temperature and high power, as well as in hard coatings. Analysis of the density of states shows that the interlayers exhibit a metallic behavior that essentially comes from the polarization of V-3d states that cross the Fermi level. The interlayers exhibit magnetic properties, with a magnetic moment of $4\mu_\beta$/cell. Due these properties, the interlayer can potentially be used in the field of spintronics.

References

[1] Beloufa A, Bensaad Z, Soudini BA, Sekal N, Bensaad A and Abid H 2009 *International Journal Nanoelectronics and Materials* **2(1)** 11
[2] Murillo J F, Ortega C, Espitia M J 2015 *Results in Physics* **5** 281
[3] Taniyasu Y and Kasu M 2008 *Diam. Relat. Mater.* **17** 1273
[4] Carcia PF, French RH, Reilly ML, Lemon MF, Jones DJ 1997 *Appl. Phys. Lett.* **70** 2371
[5] Carcia PF, French RH, Sharp K, Meth JS, Smith BW 1996 *Soc. Opt. Eng.* **2884** 255
[6] Jonnard P, Capron N, Semond F, Massies J, Martinez-Guerrero E, Mariette H 2004 *Eur. Phys. J. B.* **42(3)** 351
[7] Beheshtian J, Baei MT, Bagheri Z, Peyghan AA 2012 *Microelectron. J.* **43(7)** 452
[8] Espitia-Rico MJ, Díaz JH and Ortega C 2015 *Rev. LatinAm. Metal. Mat.* **35(2)** 165
[9] Giannozzi P, Baroni S and Bonin N 2009 *J. Phys. Condens. Matter.* **21** 395502
[10] Perdew J.P, Burke K and Ernzerhof M 1996 *Phys. Rev. Lett.* **77** 3865
[11] Vanderbilt D 1990 *Phys. Rev. B* **41** 7892
[12] Laasonen K, Pasquarello A, Car R, Lee C and Vanderbilt D1993 *Phys. Rev. B* **47** 10142
[13] Monkhorst H and Pack J 1976 *Phys. Rev. B* **13** 5188
[14] Murnaghan F.D 1944 *Proc. Natl. Acad. Sci. USA* **30** 244
[15] Vargas-Hernández C, Espitia-Rico MJ, Báez R E 2015 *computational condensed Matter* **4** 1
[16] Ortega-López C, Casiano-Jiménez G and Espitia-Rico MJ 2016 *Journal of Physics: Conference Series* **687** 012052
[17] Beloufa A, Bensaad Z, Soudini B, Sekkal N, Bensaad A, Abid H 2009 *International Journal Nanoelectronics and Materials* **2** 11
[18] Gaonzalez R, Rodríguez J, López W *Ciencia e Ingeniería Neogranadina* **17** 23
[19] Zunger A, Wei S H, Ferreira L, Bernard J E 1990 *Physical Review Letters* **65** 353

Journal of Physics: Conference Series **743** (2016) 012005 doi:10.1088/1742-6596/743/1/012005

DFT applied to the study of carbon-doped zinc-blende (cubic) GaN

M J Espitia R[1], C Ortega-López[2] and J A Rodríguez Martínez[3]
[1] Universidad Distrital Francisco José de Caldas, Bogotá, Colombia.
[2] Universidad de Córdoba, Montería, Colombia.
[3] Universidad Nacional de Colombia, Bogotá, Colombia.

E-mail: mespitiar@udistrital.edu.co

Abstract. Employing first principles within the framework of density functional theory, the structural properties, electronic structure, and magnetism of C-doped zincblende (cubic) GaN were investigated. The calculations were carried out using the pseudopotential method, employed exactly as implemented in Quantum ESPRESSO code. For $GaC_{0.0625}N_{0.9375}$ concentration, a metallic behavior was found. This metallic property comes from the hybridization and polarization of C-2p states and their neighboring N-2p and G-4p states.

1. Introduction

Gallium nitride (GaN) is one of the III-nitride semiconductors. This material has been the subject of many intense investigations over the last few decades due to potential applications for optical devices [1] and in high-voltage, high-power, and high-temperature microwave applications [2, 3]. They are wide-band-gap semiconductors and are used for short-wavelength electroluminescent devices. Additionally, heterostructures based on the surface growth of oxides on GaN are used in various applications, such as high-efficiency blue and green light-emitting devices, ultraviolet light-emitting diodes, metal-oxide-semiconductor field-effect transistors, and field-effect transistors [4–6]. The ground state of GaN is the wurtzite structure. However, depending on the substrate and the growth conditions, a zinc-blende phase can be formed [7]. The difference in total energy between the zinc-blende and wurtzite structures is very small (~10meV/atom), which explains why the two structures can be obtained experimentally. On the other hand, for a long time, carbon has been considered an atmospheric impurity that could contaminate semiconductor devices based on GaN during the manufacturing process. In particular, carbon contaminant behavior for materials such as GaAs, AlAs, and InAs grown via molecular beam epitaxy has been reported [8, 9]. However, recent theoretical studies show that carbon-doped wurtzite GaN exhibits a metallic behavior [9] and acquires magnetic properties with a magnetic moment of $0.7\mu_\beta$/cell [10]. Additionally, recently pioneering theoretical research carried out by Espitia *et al.* [11] has shown that graphene on the GaN(0001) wurtzite surface is stable and the Dirac cones remain intact. However, theoretical or experimental investigations of carbon-doped zinc-blende (cubic) GaN are scarce. For this reason, in this paper we present a systematic theoretical study of the structural, electronic, and magnetic properties of carbon-doped zinc-blende (cubic) GaN.

Content from this work may be used under the terms of the Creative Commons Attribution 3.0 licence. Any further distribution of this work must maintain attribution to the author(s) and the title of the work, journal citation and DOI.
Published under licence by IOP Publishing Ltd

Journal of Physics: Conference Series **743** (2016) 012005 doi:10.1088/1742-6596/743/1/012005

2. Computational method

The calculations were carried out within the DFT framework using the Quantum Espresso package [12]. The correlation and exchange effects of the electrons were considered using the generalized gradient approximation (GGA) of Perdew, Burke, and Ernzerhof (PBE) [13]. Electron–ion interactions were dealt with via the pseudopotential method [14, 15]. The electron wave functions were expanded into plane waves with a kinetic-energy cutoff of 40Ry. For the charge density, a kinetic energy cutoff of 400Ry was used. The calculations were performed taking into account the spin polarization due to the presence of the Cr atom. To calculate the lattice constant, the bulk modulus, and the total energy of each studied superlattice, the calculated data were fit to the Murnaghan equation of state [16]. To calculate the structural, electronic, and magnetic properties of pure GaN, a 32-atom 2ax2bx1c zincblende supercell was considered. The concentration of $GaC_{0.0625}N_{0.9375}$ was obtained by substituting for one N-atom in the zinc-blende supercell. Pure GaN and the $GaC_{0.0625}N_{0.9375}$ compounds were modeled according to the special quasirandom structures approach [17], and the disorder aspects were ignored.

3. Results and discussions

3.1. Structural properties

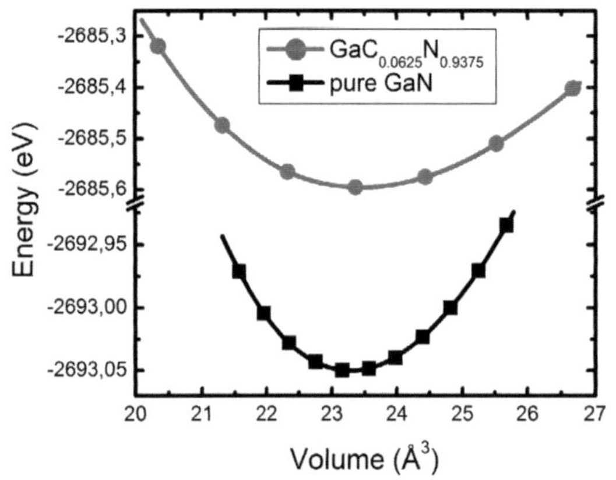

Figure 1. Total energy as a function of the volume for pure GaN (black), $GaC_{0.0625}N_{0.9375}$ (red).

In order to determine the structural properties in the ground state, such as the lattice constant (a_0), the bulk modulus (B_0), the total energy (E_0), and the magnetic moment (μ) per supercell of pure GaN and $GaC_{0.0625}N_{0.9375}$ concentration in the zincblende structure, the total energy was calculated as a function of the volume, and the results were fit to the Murnaghan equation of state [16]. It can be seen that pure GaN and $GaC_{0.0625}N_{0.9375}$ compound are stable or metastable, because in each of the energy-volume curves there is a minimum energy value.

The lattice constant, the c/a value, the bulk modulus (B_0), the total energy (E_0), and the magnetic moment (μ_β) per cell are shown in Table 1.

Table 1. Structural parameters of the lattice constant (a_0), bulk modulus (B_0), total energy (E_0), and magnetic moment (μ_β) per cell in the zinc-blende structure.

Compound	a_0 (Å)	V_0 (Å^3)	B_0 (GPa)	E_0 (eV)	μ (μ_β)
	4.533	23.258	175.75	-2693.05	0
GaN	4.552[a]	23.58[a]	175.32[a]	-	-
	4.59[b]	-	-	-	-
$GaC_{0.0625}N_{0.9375}$	4.543	23.454	171.73	-2685.59	0

[a] Theoretical reference [18], [b] Theoretical reference [19].

Journal of Physics: Conference Series **743** (2016) 012005 doi:10.1088/1742-6596/743/1/012005

We can see in the lattice constant a calculated for the pure GaN agrees well with values reported theoretically, since it differs by less than one percent. The values of the bulk modules of pure GaN and $GaC_{0.0625}N_{0.9375}$ compound are higher, which confirms that they are quite rigid, making them good candidates for possible application in devices operated at high temperature and high power, as well as in hard coatings. When one N atom in the 2ax2bx1c supercell is replaced with a C atom, the lattice constant in $GaC_{0.0625}N_{0.9375}$ compound changes only slightly with respect to pure GaN. This small change in the parameters may be due to the fact that the radius of the C atom (0.914Å) is very close to the atomic radius of N (0.92Å). A similar result has been reported by Espitia et al [9] for $GaC_{0.25}N_{0.75}$, $GaC_{0.50}N_{0.50}$, and $GaC_{0.75}N_{0.25}$ compounds in the wurtzite structure.

In order to verify the relative stability of $GaC_{0.0625}N_{0.9375}$ compound, we calculated the corresponding formation energy, which is expressed as:

$$E_f = E_{C:GaN} - E_{GaN} - nE_C + nE_N \quad (1)$$

where $E_{C:GaN}$, E_{GaN}, E_C and E_N are the energies of C-doped GaN, pure GaN, an isolated carbon atom, and a N atom in the nitrogen molecule, respectively [20]. The integer n is the number of C atoms that substitute for N. We calculated the total energy E_C and E_N in their ground states, graphite and molecule. The energy values are -11.406eV and -19.90eV, respectively. The calculated value of formation energy E_f for $GaC_{0.0625}N_{0.9375}$ is -0.751eV. The value of the formation energy of $GaC_{0.0625}N_{0.9375}$ compound is negative. Therefore, the compound is thermodynamically stable, and the moderate formation energy values indicate that the compounds can easily be grown experimentally.

3.2. Electronic properties

The theoretical lattice constants of GaN and $GaC_{0.0625}N_{0.9375}$ compound, shown in Table 1, were used to calculate the band structure and the spin-polarized density of states (DOS) along the high-symmetry direction in the first Brillouin zone.

The band structure for pure GaN and $GaC_{0.0625}N_{0.9375}$ compound are shown in Figure 2. Figure 2(a) shows the band structure of pure GaN. This confirms the direct semiconductor behavior, with the top of the valence band and the bottom of the conduction band located at the Γ point of the Brillouin zone. We found a direct band gap of about 1.78eV. The magnitude of this gap agrees well with values reported theoretically [18].

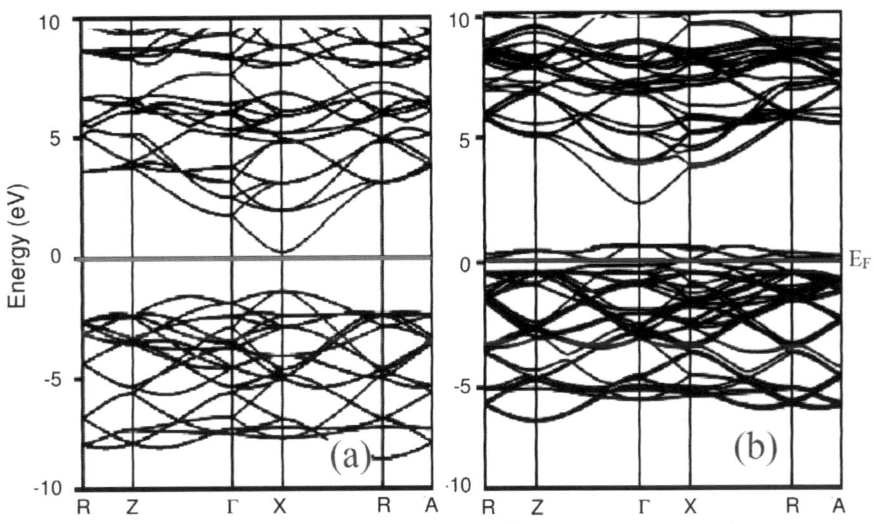

Figure 2. Band structure: (a) pure GaN, (b) $GaC_{0.0625}N_{0.9375}$ compound.

Figure 2(b) shows the band structure of the ternary $GaC_{0.0625}N_{0.9375}$ compound. We can observe that there is a penetration towards the prohibited energy zone. Therefore, due to the introduction of a C atom into the structure of GaN, it loses its semiconductor nature. The allowed ternary $GaC_{0.0625}N_{0.9375}$ compounds exhibit metallic behaviour.

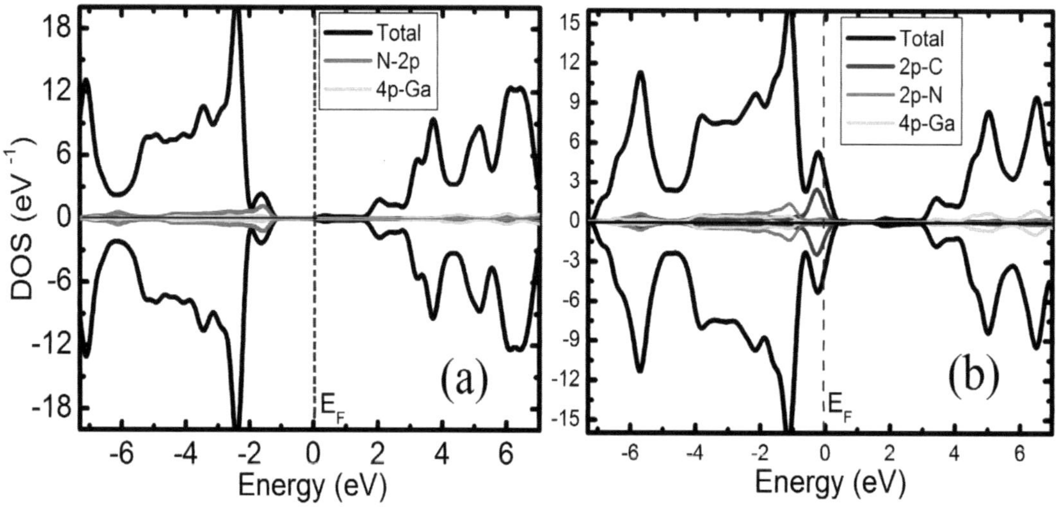

Figure 3. Partial and total density of states for the allowed ternary compounds (a) pure GaN and (b) $GaC_{0.0625}N_{0.9375}$ compound.

Figure 3(a) shows the total and partial spin-polarized density of states of pure GaN in the zinc-blende structure. The total density of states confirms that pure GaN exhibits semiconductor behavior, determined mainly by the 2p-N and 4p-Ga states. Figure 3(b) shows the total and partial spin-polarized density of states of $GaC_{0.0625}N_{0.9375}$ compound zinc-blende structure. The total density of states confirms that due to the C substitution at the N site, the compound has a metallic character. This behavior is mainly dominated by the C-2p states and to a lesser extent by the N-2p and 4p-Ga states. Similar results were reported by Yu et al. [21] in their study of carbon-doped GaN in the wurtzite structure using DFT and by Murillo et al. [22] in their study of the effects of C for N and C for Al in the w-AlN compound.

4. Conclusions

In this work, we carried out first-principles calculations in order to study the structural, electronic, and magnetic properties of pure GaN and $GaC_{0.0625}N_{0.9375}$ compound in the zinc-blende structure, within the framework of density functional theory (DFT). The calculated structural parameters agree well with values reported theoretically. The calculated values of the bulk modules of pure GaN and the $GaC_{0.0625}N_{0.9375}$ compound are quite high; therefore, these compounds are quite rigid, which makes them attractive for potential applications at high temperatures and for hard coatings. On the basis of the density of states, we found that the $GaC_{0.0625}N_{0.9375}$ compound exhibits a metallic behavior determined by the C-2p, N-2p, and Ga-4p states.

References

[1] Nakamura S 1997 *Solid State Communications* **102** 237
[2] Wagner J M and Bechstedt F 2002 *Physical Review B* **66** 115202
[3] Nakamura S, Senoh M, Iwasa N and Nagahama S 1995 *Applied Physics Letters* **67** 1868
[4] Binari S C, Kruppa W, Dietrich H B, Kelner G, Wickenden A E and Freitas J A 1997 *Solid State Electron* **41** 1549
[5] PingNAT, Chen Q, Yang J W, Asif-Khan M and Adesida I 1998 *IEEE Electron Device Letters* **19** 54

[6] Dang X Z, Asbeck P M, Yu E T, Sullivan G J, Chen M Y, McDermot B T, Boutros K S and Redwing J M 1999 *Appplied Physic Letter* **74** 3890
[7] Yeh C-Y, Lu Z W, Froyen S and Zunger A 1992 *Physics Review B* **46** 10086
[8] Ito H and Ishibashi T 1991 *J. Appl. Phys. Part 2* **30** L994
[9] Espitia M J, Moreno M G, Rodríguez J A and Takeuchi N 2016 *Journal of Physics Conference Series* **687** 012048
[10] Hu X L, Zhang YF, Zhuang N F and Li J Q 2010 *Journal of Solid State Chemistry* **183** 2741
[11] Espitia-Rico M, Rodríguez-Martínez J, Moreno-Armenta M G and Takeuchi N 2015 *Applied Surface Science* **326** 7
[12] Giannozzi P, Baroni S and Bonin N 2009 *Journal of Physics Condensed Matter* **21** 395502
[13] Perdew J, Burke K and Ernzerhof M 1996 *Physical Review Letter* **77** 3865
[14] Vanderbilt D 1990 *Physics Review B* **41** 7892
[15] Laasonen K, Pasquarello A, Car R, Lee C and Vanderbilt D 1993 *Physics Review B* **47** 10142
[16] Murnaghan F D 1944 *Proc. Natl. Acad. Sci. USA* **30** 244
[17] Zunger A, Wei S H, Ferreira L and Bernard J E 1990 *Physical Review Letters* **65** 353
[18] Arbouche O, Belgoumène B, Soudini B and Driz D 2009 *Computational Materials Science* **47** 432
[19] Gavrilenko V I and Wu R Q 2000 *Phys. Rev. B* **61** 2632
[20] Murillo J F, Ortga-López C and Espitia M J 2015 *Results in Physics* **5** 281
[21] Lin Yu L, Wang Z, Guo M, Liu D, Dai Y and Huang B 2010 *Chemical Physics Letters* **487** 251
[22] Murilo J F, Ortega C and Espitia M J 2016 *Journal of Physics Conference Series* **687** 012114

The most stable mono-layers of (111)-Pt (fcc) on Graphene: A first-principles GGA study

J Otalora-Acevedo[1,2], J A Rodríguez Martínez[1], G Moreno-Armenta[3], E Vera[2] and N Takeuchi Tan[3]
[1] Universidad Nacional de Colombia, Bogotá, Colombia.
[2] Universidad Pedagógica y Tecnológica de Colombia, Tunja, Colombia.
[3] Centro de Nanociencias y Nanotecnología, UNAM, Ensenada, B.C., México.

E-mails: jose.otalora@uptc.edu.co , jarodriguezm@unal.edu.co

Abstract. We investigate monolayers of planes (111) of Pt in the FCC structure located on graphene. The energy of formation showed that the most stable structure is $\sqrt{3}\times\sqrt{3}-$ Pt on $2\times 2-$ graphene. This system has a mismatch in the lattice constant of 0.45. The layers are completely flat, and its band structure shows that the new structure is metallic and the Dirac's cones are displaced 0.6eV above of the Fermi level. In this work we present the dependence of the enthalpy of formation of these structures and we calculated all structural parameters of their relaxation.

1. Introduction

Graphene is today a very important material because of its surprising physical and chemical properties. Many very interesting derivatives of graphene have been obtained as result of the deposition of nanoparticles or nanoclusters based on metals, semiconductors, magnetic atoms, and insulators. Today, graphene-metal particle nanocomposites [1] and Pt-graphene composites [2] are a reality. Graphene and graphene oxide (GO) are useful when they are doped or decorated with adsorbed atoms or with nanoparticles of Pt, Au, or Ag [3]. In this paper, we present a study of Pt-monolayer/graphene, with a new focus, because this system is very important in electrocatalysis [4-8]. Reference [9] reports that a 3D foam of graphene with Pt Ru exhibits very good electrochemical properties and also serves as a detector of hydrogen peroxide. The foregoing are reasons for the increasing importance of the Pt/graphene system. In order to contribute to this area of research, we carried out calculations of total energy, based on Density Functional Theory (DFT) of several structures of Pt-monolayer on graphene. From these results, we found that the most probable structure is $\sqrt{3}\times\sqrt{3}$–Pt/2×2 – graphene. This structure is metallic and exhibits the Dirac's cones.

2. Calculation details

We used Quantum-Espresso code (QE) [10] in order to carry out the total energy calculations. The GGA-PBE approximation [11, 12] was used. QE uses a base of plane waves (PWs) to expand the wave function, and we used 30Ry as the maximum of kinetic energy of the PWs in order to limit the size of the base. Likewise, QE expands the charge density in PWs, and the cutoff for the maximum of kinetic energy of the base was 320Ry. The convergence was 0.0001Ry. QE's relaxation process moves atoms in order to find the optimal positions that minimize the total energy and the forces acting on the atoms. This process ceased when the force was less than 0.001Ry/a.u. All calculations were

Content from this work may be used under the terms of the Creative Commons Attribution 3.0 licence. Any further distribution of this work must maintain attribution to the author(s) and the title of the work, journal citation and DOI.
Published under licence by IOP Publishing Ltd

Journal of Physics: Conference Series **743** (2016) 012006 doi:10.1088/1742-6596/743/1/012006

spin-polarized. Since QE works only with periodic systems, the periodicity in the direction that is perpendicular to the Pt-graphene layer was achieved by performing periodic slab calculations in which the separation between Pt-graphene bilayers was at least 12Å. In turn, this separation guarantees that the interaction between surfaces is negligible. Only one k-point was considered (Γ), because the supercells used were very large. The new approach used in this research involves the design of monolayers (111) of Pt and graphene layers. This new approach defines l_{Pt} and l_{graph} as the lengths of the sides of cells of Pt and graphene respectively, as $l_{Pt} = a_{Pt}N_{Pt}\sqrt{i}$ and $l_{graph} = a_{graph}N_{graph}\sqrt{j}$, where $a_{Pt/graph}$ are the lattice constants of 1x1 unit cells of (111) Pt and graphene, $N_{Pt/graph}$, i and j are integers. Note that $N\sqrt{i}$ is the general form of a 2D-cell. Then, we search layers such that $l_{Pt}/l_{graph} \approx$ integer or rational numbers p/q with p and q integers. Since it is impossible to get exact integers or rational numbers, a small tolerance window was selected. The tolerance chosen was 5% with respect the graphene lattice constant.

3. Results

Many structures were calculated. Table 1 shows a summary of results, and Figure 1 shows the structures as a model of spheres. The formation energy, E_f (structure), is used to compare the relative stability of structures containing different quantity of atoms, and is calculated as:

$$E_f(\text{structure}) = n_{Pt}E_{Pt}^{FCC-vol} + n_C E_C^{graphite} - E_{slab}^{DFT} \quad (1)$$

Where n is the number of Pt/C atoms in the considered structure, $E_{Pt/C}^{Fcc/graphite}$ is the energy of one atom of Pt/C in the FCC/graphite unit cell, and E_{Slab}^{DFT} is the energy of the slab calculated by QE. $E_{Pt}^{FCC-vol} = -86.6289\text{Ry}$ and $E_C^{graphite} = -11.3985\text{Ry}$.

Table 1. Detail of structures Pt/graphene proposed.

Structure		Atoms in unit cell Pt	Atoms in unit cell C	Mismatch %	Formation energy (eV) per atom
$\sqrt{3}\times\sqrt{3}$Pt/2×2graphene	Structure #3	3	8	-0.450	0.3534
	Structures #1, #2, #4, #5				0.3704
$\sqrt{7}\times\sqrt{7}$Pt/3×3 graphene	Structure #4	7	18	1.377	0.3849
	Structures #1, #2, #5, #8, #9				0.3873
	Structures #3, #6, #7				0.3874
3×3Pt/2$\sqrt{3}$×2$\sqrt{3}$graphene	Structure #3	9	24	-0.450	0.3746
	Structure #1				0.3747
$\sqrt{13}\times\sqrt{13}$Pt/4×4graphene	Structure #1	13	32	3.615	0.4384
4×4Pt/$\sqrt{21}\times\sqrt{21}$graphene	Structure #6	16	42	0.337	0.3751
	Structures #7, #3				0.3752
	Structure #2				0.3755
	Structure #4				0.3762

In Table 1, column 2, different structures correspond to different relative positions of the graphene layer with respect to the layer of *Pt*, which were obtained by relative displacement. Therefore, *C* atoms are placed at different positions on top the unit cell of (111)-Pt layer. The structure with the smallest formation energy is $\sqrt{3}\times\sqrt{3}$Pt /2×2graphene structure #3 (Pt atoms at two different bridge points of the graphene), which is shown in Figure 2(a). The mismatch is calculated as:

$$\text{mismatch} = \frac{l_{Pt-cel} \times a^{attice}_{(111)Pt}}{l_{graphene} \times a^{lattice}_{grahene}} - 1 \tag{2}$$

With: $a^{lattice}_{graphene} = 2.4595\text{Å}$, $a^{lattice}_{(111)Pt} = 2.8272\text{Å}$, and l is the length of the cell of *Pt* or *graphene* of column 1.

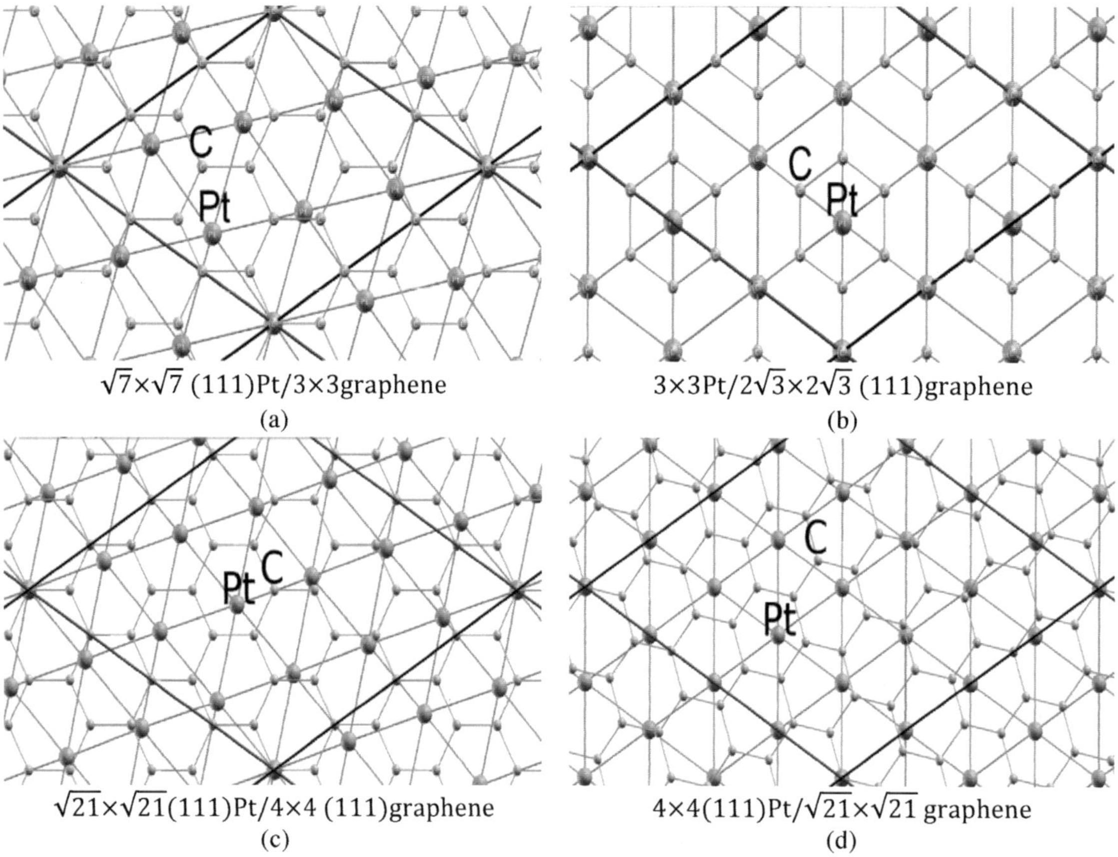

Figure 1. Top view of the structures (111) Pt /graphene, mentioned in Table 1.

Figure 2(b) shows that both layers are completely flat. No deformation is induced in the graphene or Pt layers. The separation between the layers is about 4.14Å, and it seems that the interaction in the vertical direction is weak. However, the electronic bands show that the influence of the Pt atoms on the graphene is sufficient to produce an alteration in the bands with respect to bands of the isolated graphene. Figure 2(c) shows the bands of this structure. This structure exhibits no magnetic moment. Although the semi-metallic behavior of graphene is lost, the Dirac's cones have moved toward the conduction band by about 0.6eV. This feature is emphasized in Figure 2(c) with a red circle that highlights the cones. Other derivative structures are obtained by means of relative displacements between the layers.

Journal of Physics: Conference Series **743** (2016) 012006 doi:10.1088/1742-6596/743/1/012006

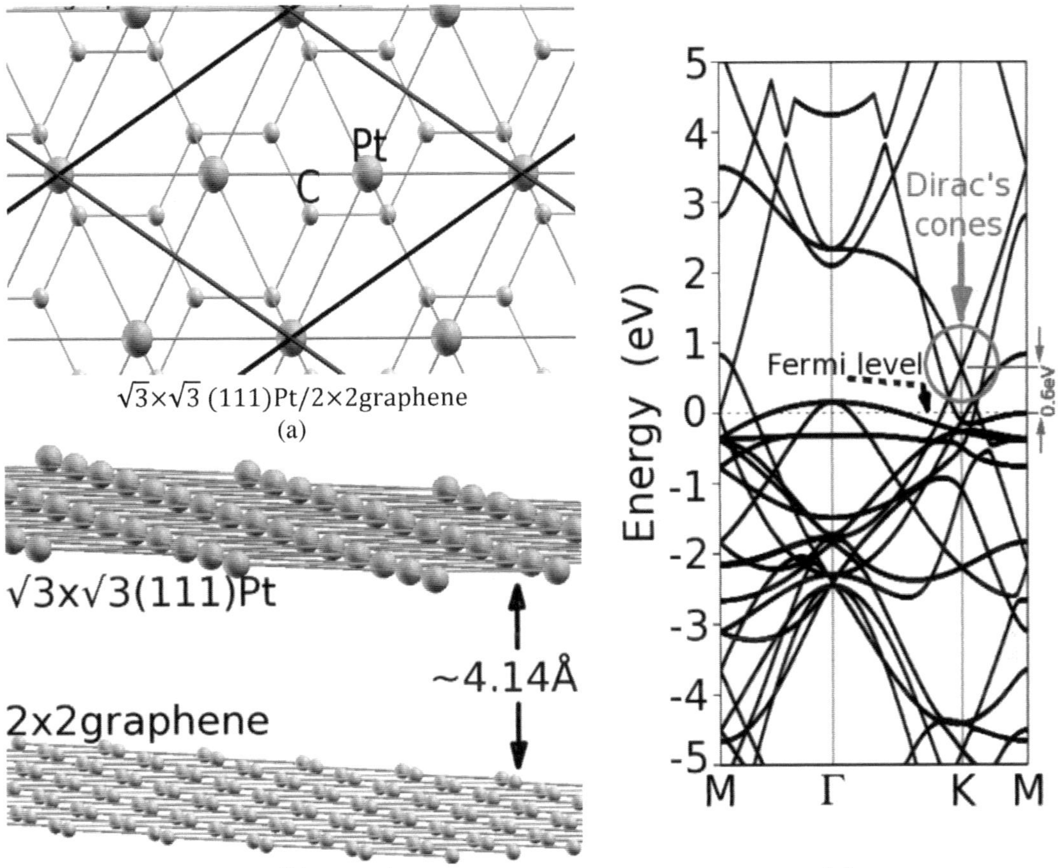

(b) (c)

Figure 2. (a) $\sqrt{3}\times\sqrt{3}-$ Pt/ $2\times2-$ graphene. Black lines show the limits of the unit cell of the structure. As can be seen, this lattice constant contains two graphene lattice constants and $\sqrt{3}$ times the lattice constant of the surface (111) of Pt in the FCC structure. Additionally, it can be seen that the Pt atoms are situated at the two bridge points over the graphene. (b) Lateral view of the structure. (c) The band structure of the stable structure. The Dirac's cones are highlighted with a red circle.

4. Summary and conclusions

We presented calculations of total energy of systems composed of one monolayer of (111) of Pt in the FCC structure located on graphene. The structure with the smallest formation energy is $\sqrt{3}\times\sqrt{3}-$ Pt/ $2\times2-$ graphene, the separation between the layers is about 4.14Å, and the layers are flat. On the other hand, the bands show that it exhibits metallic behavior, its magnetic moment is zero, and the Dirac's cones are located 0.6 eV above the Fermi level.

Acknowledgments

Calculations were carried out in the DGCTIC-UNAM under projects SC16-1-IR-57. MGMA thanks the DGAPA-UNAM project IN102714-3 in Mexico and in the cluster of the GEMA Group at the Physics Dept. of Universidad Nacional de Colombia (UN), Bogotá. J. Otálora was partially supported by the Doctoral Program of Materials of Universidad Pedagógica y Tecnológica, Tunja, Boyacá, Colombia.

References

[1] Chao Xu, Xin Wang and Junwu Zhu 2008 Graphene-Metal Particle Nanocomposites *J. Phys. Chem. C* **112** 19841–19845

[2] Yongchao Si and Edward T Samulski 2008 Exfoliated Graphene Separated by Platinum Nanoparticles *Chem. Mater.* **20** 6792–6797

[3] G Giovannetti, P A Khomyakov, G Brocks, V M Karpan, J van den Brink and P J Kelly 2008 Doping Graphene with Metal Contacts *Phys. Rev. Lett.* **101** 026803

[4] EunJoo Yoo, Tatsuhiro Okata, Tornoki Akita, Masanori Kohyama, Junji Nakamura and Itaru Honma 2009 Enhanced Electrocatalytic Activity of Pt Subnanoclusters *Graphene Nanosheet Surface* **9(6)** 2255-2259

[5] Brian Seger and Prashant V Kamat 2009 Electrocatalytically Active Graphene-Platinum Nanocomposites. Role of 2-D Carbon Support in PEM Fuel Cells *J. Phys. Chem. Lett. C.* **113** 7990–7995

[6] Shaojun Guo, Shaojun Dong and Erkang Wang 2010 Three-Dimensional Pt-on-Pd Bimetallic Nanodendrites Supported on Graphene Nanosheet: Facile Synthesis and Used as an Advanced Nanoelectrocatalyst for Methanol Oxidation *ACS Nano* **4(1)** 547-555

[7] Rong Kou, Yuyan Shao, Donghai Wang, Mark H. Engelhard, Ja Hun Kwak, Jun Wang, et al. 2009 Enhanced activity and stability of Pt catalysts on functionalized graphene sheets for electrocatalytic oxygen reduction *Electrochemistry Communications* **11** 954–957 doi: 10.1016/j.elecom.2009.02.033

[8] Yongjie Li, Wei Gao, Lijie Ci, Chunming Wang, Pulickel M. Ajayan 2010 Catalytic performance of Pt nanoparticles on reduce graphene oxide for methanol electro-oxidation *Carbon* **48** 1124-1130 doi: 10.1016/j.carbon.2009.11.034

[9] Chih-Chien Kung, Po-Yuan Lin, Frederick John Buse, Yuhua Xue, Xiong Yu, Liming Dai, Chung-Chiun Liu 2014 Preparation and characterization of three dimensional graphene foam supported platinum–ruthenium bimetallic nanocatalysts for hydrogen peroxide based electrochemical biosensors *Biosensors and Bioelectronics* **52** 1–7 doi: 10.1016/j.bios.2013.08.025

[10] P Giannozzi, S. Baroni, N Bonini, M Calandra, R Car, C Cavazzoni, D Ceresoli, G L Chiarotti, M Cococcioni, I Dabo, A. et. al. 2009 *J. Phys. Condens. Matter.* **21** 395502

[11] *Pt.pbe-nd-rrkjus.UPF* http://www.quantum-espresso.org/wp-content/uploads/upf_files/Pt.pbe-nd-rrkjus.UPF

[12] John P Perdew, Kieron Burke and Matthias Ernzerhof 1996 *Phys. Rev. Lett.* **77(18)** 3865-3868

Journal of Physics: Conference Series **743** (2016) 012007 doi:10.1088/1742-6596/743/1/012007

First-principles study on the formation energies of $Ga_{1-x}Cr_xAs$

N Martínez[1,2] and J A Rodríguez Martínez[1]
[1] Universidad Nacional de Colombia, Bogotá. Colombia.
[2] Universidad de los Llanos, Villavicencio, Colombia.

E-mail: nmartinez@unillanos.edu.co

Abstract. The electronic structure and formation energy of the $Ga_{1-x}Cr_xAs$ compound has been studied by principles of methods based on the Density Functional Theory. The formation energy of the compound is calculated, as well as the vacancies and substitutions. The equations of Kohn-Sham are solved employing the plane wave method and the ultra-soft atomic pseudopotentials approximation. The generalized gradient approximation was employed for the exchange and correlation energy in the Perdew-Burke-Ernzerhof (PBE) parametrization as it is implemented in the computational code Quantum Espresso. It was found that the most favourable configuration is the GaAs lattice by doing the substitution of one Ga atom by one Cr in the analyzed lattices.

1. Introduction

The analyzed structures in this study of the formation energy compound $Ga_{1-x}Cr_xAs$ are: cubic cell of pure GaAs, 1x1x2 and 2x2x2 supercells, with vacancies and substitutions as detailed below.

The $Ga_{1-x}Cr_xAs$ (x=0.25) structure is presented in the Figure 1, the substitution was done in the position (0.0, 0.5, 0.5). As each atom is bound to other 4 atoms of opposite kind, forming a regular tetrahedron, the surrounding of each one is similar. For instance, it can be done in any position of a Ga atom and the results will not present relevant variations because of the lattice symmetry. When a Ga atom is substituted by one Cr in the right position, the structure is considered cubic.

For each As atom the surrounding is similar, since it is bound to 4 Ga atoms, forming a regular tetrahedron. A similar situation happens for each Ga atom that bonds with four As atoms. The As atom that is located at position (0.25, 0.25, 0.25) is bound to the Ga atom that is located at (0.0, 0.5, 0.5). At the Ga atom site, the substitution by the Cr atom was made; and there, the Ga vacancy was made too. As a result, the As vacancy was generated at (0.25, 0.25, 0.25). This process was necessary to do the comparisons and analysis of the structural, electronic and magnetic alterations of the compounds.

In order to analyse the possible alterations that the formation energy of the $Ga_{1-x}Cr_xAs$ suffers, the concentration of the Cr atom was varied from 0.25 to 0.125, creating a 1x1x2 structure. The concentration value of the Cr atom agrees with what is reported by L. M. Sandratskii *et al* [1] and Masafumi Shirai [2]. The substitutions of the Ga atom were made by the Cr in different positions, finding that the most stable is at (0.50, 0.50, 0.00). In the same way, it was demonstrated that in terms of formation energy, it will not change significantly in the diverse positions considered for the Cr atom. In the structure without vacancies and no substitutions, the As atom is at position (0.75, 0.75, 0.125). It is bound to four Ga atoms; one of them is located at (0.50, 0.50, 0.00). It was made a substitution by Cr, in the mentioned site of the Ga. In addition, the Ga vacancy was made in that site

Content from this work may be used under the terms of the Creative Commons Attribution 3.0 licence. Any further distribution of this work must maintain attribution to the author(s) and the title of the work, journal citation and DOI.
Published under licence by IOP Publishing Ltd

also. Due to the reasons mentioned before, the As vacancy was generated in the site indicated at the beginning of this paragraph.

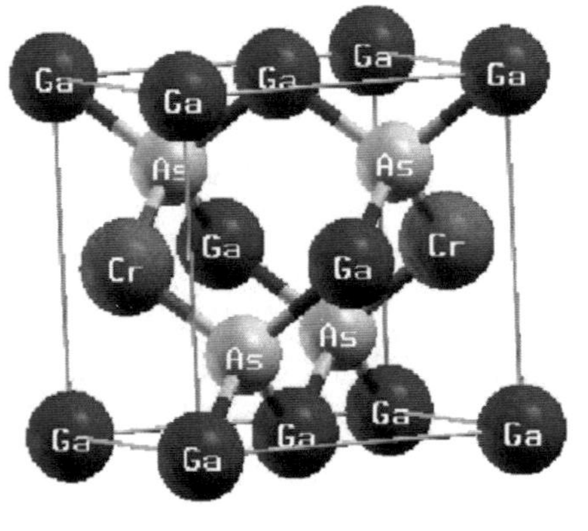

Figure 1. The crystal structure of Ga(Cr)As.

The 2x2x2 supercell has a lattice constant a=11.32Å. The Zincblende structure was taken to calculate it, and it was repeated in each axis twice. When the formation energy of the compounds is analysed, as presented in the results chapter, it is found that changing the atomic position for the substitution; does not exhibit big alterations. It is due to the reason that the surrounding of each As atom is similar, as well as the surrounding of each Ga atom. In addition, a 2x2x2 supercell was employed for the compound $Ga_{1-x}Cr_xAs$ with a Cr concentration of 0.03125. The concentration value of the dopant agrees with what is reported by V.I. Baykov, *et al* [3], and L. M. Sandratskii *et al* [1]. A substitution of one Ga atom by one Cr was made in this 2x2x2 supercell, at position (0.00, 0.75, 0.25), because of the reasons explained in the previous paragraph.

In order to do the comparisons and analysis of the structural, electronic and magnetic alterations of the compounds, it was necessary to do a substitution of the Ga atom at position (0.00, 0.75, 0.25) by a Cr atom. That Gallium atom is bound to other four As atoms, one of them is located at position (0.125, 0.125, 0.125). In that site, an As vacancy was made, and a Ga vacancy was also made in its corresponding site.

2. Parameters optimization

The calculations were done under the formalism of the Density Functional Theory, employing the plane wave methods and the ultrasoft atomic pseudopotentials, as it is implemented in the computational code Quantum Espresso [4]. The plane wave basis was expanded to an energy cut-off of 30Ry. The G vectors that allow the expansion are vectors of reciprocal lattice. Initially, an infinite number of vectors are required to represent the wave functions with an infinite precision. However, the expansion of the plane waves may be truncated to a finite number of terms, to limit the wave functions with a lower kinetic energy than an energy cut-off E_{cutoff} [5].

The Generalized Gradient Approximation (GGA) was used to get the exchange and correlation effects, in the parametrization of Perdew, Burke and Ernzerhof (PBE) [6]. The GaAs was studied in the Zincblende crystallization phase, with a constant optimized lattice of 5.66Å, employing a k Monkhorst- Pack point mesh [7] of 6x6x6, in the first Brillouin zone.

In order to choose the entrance parameters of the calculations, the Energy cut-off was optimized and the K points mesh, according to the equilibrium volume and the lattice constant. The value of the minimum energy is calculated by using the Murnaghan state equation, equation [8].

3. Results and discussion

In Table 1, it is presented a comparison of the value of the lattice constant and the calculated volume module, for GaAs in the zincblende structure with other data reported in the literature. Error rates relate to each one in relation to the parameters chosen.

Table 1. Lattice constant, bulk modulus, calculation method, the error rate for the GaAs.

a_o (A)	B_o (GPa)	Method	%a_o (A)	%B_o (GPa)
5.608 [9]	75.2 [9]	FP-LDA	0.94%	-18.94%
5.748 [9]	60.8 [9]	FP–GGA	-1.51%	0.24%
5.654 [10]	77.0 [11]	Experiment	0.12%	-20.84%
5.653 [12]	76.0 [12]	Experiment	0.14%	-19.80%
5.66*	60.95*	Pseudopotentials		

*Present work

The values reported in reference [9] are theoretical, calculated by the method FP-FP-LDA, and GGA Full-potential linear augmented plane wave (FP-LAPW) with local density approximation -LDA-, and generalized gradient approximation (GGA), respectively. The calculations were done by the WIEN2K code; the data reported by R.W.G. Wyckoff [10], K. H. Hellwege, [11] and B. Fillipi, *et al* [12] is experimental. The difference among the calculated values is due to the methods used to calculate that parameter.

In order to calculate the most stable structure, the formation energy was calculated using the following expression [13]:

$$E_f=E_b^{doped}+\mu_y-E_b^{o}+\mu_x \tag{1}$$

Where E_b^{doped} and E_b^{o} are the total energies of the supercell with and without defects respectively, and μ_y, μ_x are chemic potential of the atom that was substituted and of the impurity that in this case is from the Cr. The formation energy is presented in Table 2.

Regarding the GaAs formation energy, it is found that the calculated value agrees with the results reported in the review of literature [14].

According to the data in the Table 2, and represented in Figure 2, the structures with less formation energy are: GaAs substituting a Gallium atom in the lattice by one Chromium, GaAs with a As vacancy, and GaAs with a Ga vacancy as it can be observed.

Table 2. Energy of formation for the compounds studied for GaAs in zincblende structure, and the *1x1x2* and *2x2x2* supercell.

Structure	*1x1x1* E_f (eV)	*1x1x2* E_f (eV)	*2x2x2* E_f (eV)
GaAs	-0.7	-0.7	-0.7
GaAs V_{Ga}	3.01	3.15	3.16
GaAs V_{as}	2.55	2.92	3.18
GaAs Ga_{Cr}	1.59	1.66	1.73
GaAs As_{Cr}	3.88	4.02	4.21
GaAs Ga_{Cr} V_{Ga}	3.92	4.44	4.63
GaAs Ga_{Cr} V_{As}	3.57	3.85	4.17

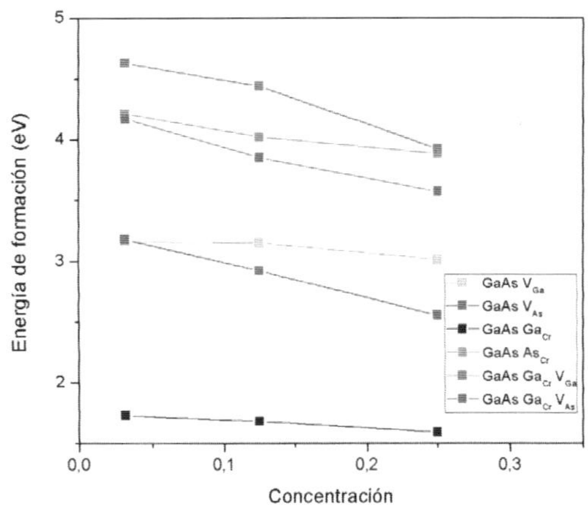

Figure 2. Energy of formation as a function of the concentration of the impurity.

4. Conclusions

According to the results, when the formation energy was analysed for the GaAs structures in the 1x1x1unit cell, 1x1x2, and 2x2x2 supercells, in their pure structure, with vacancies and substitutions, it was found that the lattice constant calculated by pseudopotentials is in agreement with the data reported experimentally. The most favourable energy-efficient configuration or the most favourable energetically is the GaAs, when it is made a substitution of one Ga atom by one Cr in the lattice, for the lattices 1x1x1, 1x1x2, and 2x2x2. The least favourable structure energetically is the GaAs making simultaneously a substitution of a Ga atom by a Cr one and a As vacancy. A projection of this study is to consider bigger supercells, it is to work with fewer concentrations of the dopants, since when the size of the supercell increases, the formation energy starts to diminish.

References

[1] L M Sandratskii and P Bruno 2003 Electronic structure, exchange interactions, and Curie temperature in diluted III-V magnetic semiconductors: (GaCr)As, (GaMn)As, (GaFe)As *Physical Review B* **67** 214402

[2] Masafumi Shirai 2001 Electronic and magnetic properties of 3d transition-metal-doped GaAs *Physica E* **10** 143-147

[3] Baykov V I, Korzhavyi P A, Smirnova E A, Abrikosov I A and Johansson B 2007 Magnetic properties of 3d impurities in GaAs *Journal of Magnetism, Magnetic Materials* **310(3)** 2120–2122

[4] P Giannozzi, S Baroni, N Bonini, M Calandra, R Car, C Cavazzoni, D Ceresoli, G L Chiarotti, M Cococcioni, I Dabo, A D Corso, S de Gironcoli, S Fabris, G Fratesi, R Gebauer, U Gerstmann, C Gougoussis, A Kokalj, M Lazzeri, L Martin-Samos, N Marzari, en Francesco Mauri, R Mazzarello, S Paolini, A Pasquarello, L Paulatto, C Sbraccia, S Scandolo, G Sclauzero, A P Seitsonen, A Smogunov, P Umari and R M Wentzcovitch 2009 Quantum espresso: a modular and open-source software project for quantum simulations of materials *Journal of Physics Condensed Matter* **21(39)** 19

[5] J Kohanoff 2006 *Electronic Structure calculations for solids and molecules: Theory and computational methods* (United Kingdom: Cambridge University Press) p 382

[6] J P Perdew, K Burke and M Emzerhof 1996 *Physical Review Letters* **77** 3856

[7] H J Monkhorst and J D Pack 1976 Special points for brillouin-zone integrations *Physical Review B* **13** 5188-5192

[8] F D Murnaghan 1944 The Compressibility of Media under extreme Pressures *Proceeding of National Academy of Sciences of the United States of America* **30(9)** 244-247

[9] R Ahmed, S J Hashemifar, H Akbarzadeh, M Ahmed and F E Aleem 2007 Ab initio study of structural and electronic properties of iii-arsenide binary compounds *Computational Materials Science* **39** 580-586

[10] Ralph W G Wyckoff f 1986 *Crystal Structures 2nd edition* (Malabar: Krieger Publishing) p 588

[11] K Hellwege and O Madelung 1982 *Semiconductors Physics of Group IV Elements and III-V Compounds New Series Group III* vol 17a, ed Landolt Börnstein (Berlin: Springer-Verlag)

[12] B. Fillipi, D.J. Singh and C.J. Umrigar 1994 *Physical Review B* **50** 14947

[13] J A Yan, C Y Wang and S Y Wang 2004 Energetics, electronic structure and local magnetism of single 3d impurity in gaas *Physics Letters A* **324** 247-253

[14] P Mahadevan and A Zunger 2003 Ferromagnetism in Mn-doped GaAs due to substitutional interstitial complexes *Physical Review B* **68** 075202

Journal of Physics: Conference Series **743** (2016) 012008 doi:10.1088/1742-6596/743/1/012008

Magnetoexciton in nanotube under external electric field

L F Garcia Russi[1], H Paredes Gutiérrez[1], Y F Santos[1] and I D Mikhailov[1]
[1] Universidad Industrial de Santander, Bucaramanga, Colombia.

Email: hparedes@uis.edu.co

Abstract. We study the Aharonov-Bohm oscillation of the energy levels of an electron-hole pair confined in a narrow nanotube in the presence of the magnetic field applied along the symmetry axis. We show that the electric field applied at the same direction makes the oscillation more pronounced.

1. Introduction

The Aharonov-Bohm (AB) effect of an exciton confined in a ring-like nanostructure, related to the interaction of the external magnetic field with the induced tunnelling current through the electron-to-hole rotation, has been studied both experimentally and theoretically [1, 2]. At the same time the theoretical analysis showed that tunnelling through potential barriers in electron-hole relative rotation is very feeble and there is no AB oscillation in ground state energy of the exciton [3, 4]. Yet, AB effect in a ring-like structure can be enhanced when the exciton is polarized in the radial or axial directions [5]. A quantum nanotube (NT), where the exciton could be polarized additionally at the axial direction by the application of an external electric field, is a possible structure for observing a stable AB effect of an electron-hole composite system. The sophisticated techniques of crystal growth developed recently permit to fabricate narrow NTs and to observe the oscillation of photoluminescence peak energies due to the AB effect of a charged exciton in a quantum tube [6, 7].

Above we present a theoretical analysis of spectral properties of the electron-hole pair confined in a very thin ideally cylindrical NT structure in the presence of electric and magnetic fields, applied along the symmetry axis. We show that the AB effect is strongly affected by the external electric field. Our calculations show that the external electric field increases the separation between electron and hole in a narrow NT, restoring an independent rotation of the particle at opposite directions and making possible the AB oscillation of the ground state level, especially for quantum NTs with a large radius.

2. Theoretical model

We consider a model of an endlessly long and very thin cylindrical nanotube, whose thickness w is much smaller than the centreline radius, in accordance to dimensions of experimentally fabricated tubes are essentially smaller than their radius [8]. Moreover, this condition allows us to reduce the number of variables in the Schrödinger equation describing in this case the electron and hole positions by means of only four cylindrical coordinates $-\pi \le \theta_e, \vartheta_h < \pi;\ -\infty < z_e, z_h < +\infty$, neglecting the particles displacements at radial direction ($\rho_e \approx \rho_h \approx R$). By using the centre-of-mass coordinates,

$$Q = \frac{m_e^* q_e + m_h^* q_h}{m_e^* + m_h^*};\ q = q_e - q_h;\ Z = \frac{m_e^* z_e + m_h^* z_h}{m_e^* + m_h^*};\ z = z_e - z_h \qquad (1)$$

In this coordinates the dimensionless electron-hole Hamiltonian acquires the following form:

Content from this work may be used under the terms of the Creative Commons Attribution 3.0 licence. Any further distribution of this work must maintain attribution to the author(s) and the title of the work, journal citation and DOI.
Published under licence by IOP Publishing Ltd

$$H = H_{cm} + H_{rel};\ H_{cm} = -\frac{\mu}{(m_e + m_h)}\left[\frac{1}{R^2}\frac{\partial^2}{\partial\Theta^2} + \frac{\partial^2}{\partial Z^2}\right];\ H_{rel} - \frac{\partial^2}{\partial z^2} + \left(\frac{i}{R}\frac{\partial}{\partial\theta} + \frac{\gamma R}{2}\right)^2 - \beta\cdot z + V_{eh}(z,\theta) \quad (2)$$

Here we use the effective atomic units, in which the effective Bohr radius $a_0^* = {}^2\varepsilon/\mu e^2$, the effective Rydberg $R_y^* = e^2/2\varepsilon a_0^*$, parameters $\gamma = e\ B/2\mu c R_y^*$ and $b = ea_0^*/Ry^*$ are used as units of length, energy and the dimensionless magnetic and electric fields, being the electron, hole and reduced effective masses, respectively m_e^*, m_h^* and $\mu = m_h^* m_e^*/(m_h^* + m_e^*)$. In our calculations we adopt a simple model, in which the confinement potential is supposed to be equal to zero inside a narrow NT region and to infinity otherwise, while the energy of the electron-hole interaction can be approximated by the expression [9]:

$$V_{eh}(z,\theta) \approx -2\Big/\sqrt{z^2 + 4R^2\sin^2\theta/2 + (0.2w)^2};\ z = z_e - z_h;\ \theta = \theta_e - \theta_h \quad (3)$$

Where w thickness is the thickness the layer. Eigenvalues of the Hamiltonian (2), $E_X = E_{N,M}^{cm} + E_{n,m}^{rel}$ corresponding to the exciton energies depend on four quantum numbers, the axial centre of mass and relative quantum numbers $N, n = 1,2,\ldots$ and corresponding quantum numbers of angular momenta ($M, m = 0, \pm 1, \pm 2, \ldots$). The centre of mass energy can be found explicitly:

$$E_{N,M}^{cm} = \frac{\mu}{(m_e + m_h)}\left[M^2/R^2 + \pi^2 N^2/L^2\right];\ N = 1,2,\quad ;M = 0,\pm 1,\pm 2, \quad (4)$$

On the contrary, the energy of the relative motion $E_{n,m}^{rel}$ can be found only numerically. But, the numerical procedure may be simplified taking advantage of the adiabatic approximation based on the fact that the rotation around the axis is significantly faster than the displacement along the z axis. In framework of this approximation, the eigenfunctions $\psi_{n,m}^{rel} = u_n(z,\theta)\cdot\phi_{m,n}(\theta)$ and eigenenergies $E_{n,m}^{rel}$ of the Hamiltonian H_{rel} can be found by using the following two-step numerical procedure:

$$-\frac{d^2u_n(z,\theta)}{dz^2} + V_{eff}(z,\theta)u_n(z,\theta) = E_n^{fast}(\theta)u_n(z,\theta);\ V_{eff}(z,\theta) = -\beta\cdot z + V_{eh}(z,\theta) \quad (5a)$$

$$\left(i\frac{1}{R}\frac{d}{d\theta} - \frac{\gamma R}{2}\right)^2\phi_{m,n}(\theta) + E_n^{fast}(\theta)\cdot\phi_{m,n}(\theta) = E_{n,m}^{rel}\cdot\phi_{m,n}(\theta) \quad (5b)$$

In our numerical work we first solve Equation (5a) by using the trigonometric sweep method [9] in order to find the adiabatic potential $E_n^{fast}(\theta)$ for 181 different angles $\theta_i = \pi(i-1)/180,\ i = 1,2,\ldots,181$ and for two lower axial quantum numbers $n = 1,2$. Afterwards, we use the found values of the function $E_n^{fast}(\theta)$ in order to solve the eigenvalue problem (5b) via Fourier series expansion method. The periodic shape of the adiabatic potential $E_n^{fast}(\theta) = E_n^{fast}(\theta + 2\pi)$ allows us to analyse qualitatively properties of the solutions of the Equation (5b) by using the Bloch's theorem. Really, via the substitution $\phi_{m,n}(\theta) = \exp(i\gamma R^2\theta/2)\psi_n(\theta)$, Equation (5b) is reduced to 1D Schrödinger equation,

$$-\frac{1}{R^2}\frac{d^2\psi_n(\theta)}{d\theta^2}+E_n^{fast}(\theta)\cdot\psi_n(\theta)=E_{n,m}^{rel}\cdot\psi_n(\theta) \tag{6}$$

As the adiabatic potential has the period 2π, Equation (6) has formally the Bloch-type solutions $\psi(\theta)=e^{ik\theta}\chi_{k,\gamma}(\theta)$ with $0<k\le 1$ and a periodic function $\chi_{k,\gamma}(\theta)$ with the same period.

3. Results and discussion

Below, we present results of calculations, obtained by numerical solution of differential Equations (5a) and (5b) for NTs with the height $L=10a_0{}^*$, the width $w=0.4a_0{}^*$, and two different centreline radius $R = 1a_0^*$, and $R = 4a_0^*$. In our calculations we use typical for GaAs material parameters, effective Bohr radius 10nm, effective Rydberg 6meV and dimensionless magnetic field $\gamma=1$ corresponds approximately to $B\approx 7T$. In Figure 1 we present results of calculation for zero electric field case. At the left side we display the adiabatic potentials found from Equation (5a) for states with axial quantum $n=1$ and $n=2$, marked by solid and dashed lines, respectively. Additionally, by horizontal lines are indicated the edges of two lowers subbands in NT of radius and three lowers subbands in NT of radius $4a_0^*$. Comparing curves in Figures 1(a) and (c) one can see that the barriers between periodical potential wells become higher and the separation between them becomes larger as the NT radius increases from $1a_0^*$ to $4a_0^*$. Consequently, in a NT of large radius the electron-to-hole tunnelling through barriers is weaker and subbands are narrower than those in a NT of small radius.

At right side of Figure 1, we display some lower energy $E_{n,m}^{rel}(\gamma)$ as functions of the magnetic field found from Equation (5b) for states with axial quantum numbers $n=1$, (solid lines) and $n=2$ (dashed lines). The meaning of these curves as we point out above is related to the fact that their shapes within the interval $0<\gamma\le 2/R^2$ coincide with the dispersion curves of the energy bands corresponding to electron-hole relative rotation around the symmetry axis. It is seen that these curves exhibit oscillations with the period $T=2/R^2$ and crossovers of the energy levels, typical for AB effect beginning from the second level in NT of radius $1a_0^*$ and beginning from the third level in NT of radius $4a_0^*$. There is an only one localized energy level N_1 in NT of radius $1a_0^*$ and there are three localized levels N_1, N_2, N_3 in NT of radius $4a_0^*$, which are arranged below the set of superior intersected curves.

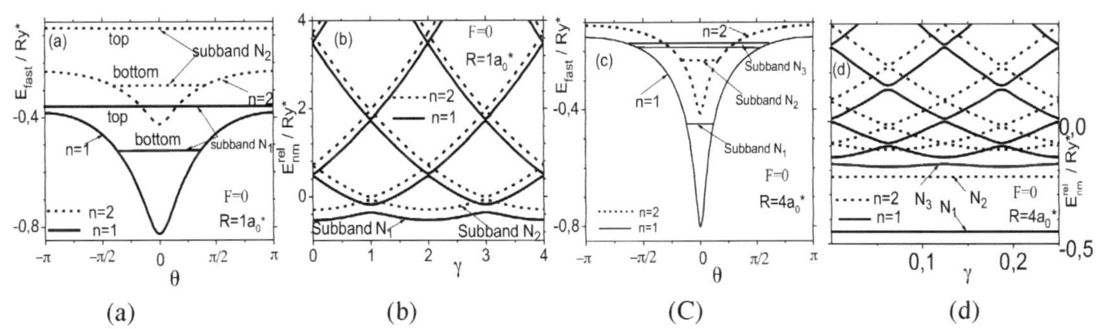

(a) (b) (C) (d)

Figure 1. (a), (c) Adiabatic periodical potentials for relative electron-hole rotation and (b), (d) lower energies of the electron-hole pair as functions of the magnetic field for zero-case electric field and for two different axial quantum numbers, $n=1$ (solid lines) and $n=2$ (dash lines).

The deeper the energy level in Figures 1(a) and (c), the narrower is the correspondent subband and larger is the amplitudes of oscillations in Figures 1(b) and (d). In the case of the small NT radius for both lower N_1 and N_2 are located only partially below the barrier top, therefore the tunnelling in these states is strong and the amplitude of the AB oscillation is large. Conversely, for NT of radius $4a_0^*$ both lower subbands N_1 and N_2 in Figure 1(c) are located completely below the barrier top, the tunnelling is very weak and the amplitude of the AB oscillations in the corresponding curves of Figure 1(d) is almost zero.

The comparison of two lower curves N_1 and N_2 in Figures 1(c) and (d) show clearly that an increase of the NT radius yields a successive quenching of their AB oscillations, related to transition of the electron-hole complex to a Wigner-molecule-like state with practically frozen relative rotation [10]. This transition is resulted from a competition between the kinetic energy $\left(K \propto 1/R^2\right)$, assisting to tunnelling and the Coulomb attraction $(V \propto -1/R)$, which impedes the tunnelling. For NTs of large radius the Coulomb attraction becomes dominating and an almost independent electron-to-hole gas-like rotation, typical for NTs of small radius, is transformed into oscillation of a structure with a molecule-like arrangement. The independence of the energy levels N_1 and N_2 in Figure 1(d) on the magnetic field proves that their configurations are similar to Wigner molecule, for which the tunnelling in these states is very weak.

Such transition should be reflected in the fluorescence spectrum of NTs, presenting a drastic change in a structuring and a blue shift of the spectral lines with increase of the NT radius. A specific interest of this effect is related to a possibility to control the amplitude of the oscillation of the energy levels by means of application of the external electric field along the symmetry axis. In the presence of a strong external electric field the separation between electron and hole grows, the energy of the attraction between them is diminished making instable the electron-hole the bound state.

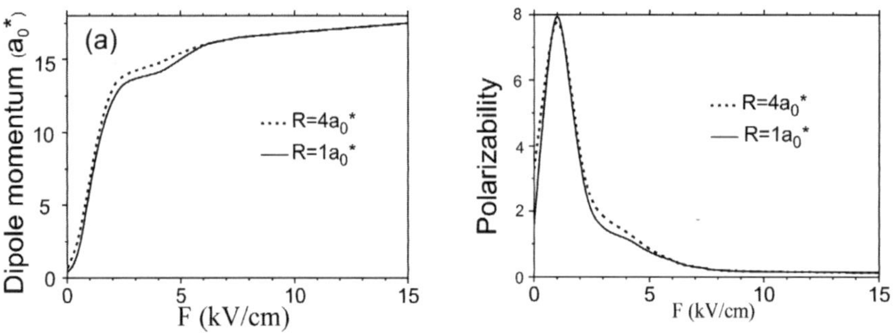

Figure 2. Dipole moment and polarizability induced by exciton confined in nanotubes of radius $R = 1a_0^*$ (solid lines) and $R = 4a_0^*$ (dash lines) as functions of the electric field applied along the symmetry axis.

We have performed similar calculations for magnetoexcitons confined in NTs in the presence of the external electric field, applied along the symmetry axis. In order to give a simple explanation of the effect of the electric field on magnetoexciton energies let us first to analyse induced by it the dipole moment and the polarizability. The dipole moment of electron-hole pair $p_{n,m}$ in the state with the relative energy $E_{n,m}^{rel}$ one can calculate as $p_{n,m} = dE_{n,m}^{rel}/dF$ and the polarizability as $\alpha_{n,m} = dp_{n,m}/dF$. In our calculations we found that dependencies of both the dipole moment and the polarizability on the electric field are very similar for different values of the magnetic field and different energy states. Therefore, in Figure 2 we present as examples of such dependencies for NTs of radius $R = 1a_0^*$ and $R = 4a_0^*$, only for the ground state $(n = 1, m = 0)$ and the magnetic field $\gamma = 0$.

One can see that the separation between the electron and the hole and the polarization of the exciton induced by the electric field are increased successively while electric field grows up the value about 2kV/cm and afterwards, the dipole moment of magnetoexcitons in NTs of radius becomes almost unchanging while the polarizability tends to zero. According to this estimation, the electron-hole separation in NTs of radius $R = 1a_0^*$ and $R = 4a_0^*$ under electric field $\varepsilon = 15\,kV/cm$, for which we below in Figure 3 present results of calculations, becomes such large that their relative movement can be considered as almost independent.

At left side of Figure 3 we display the adiabatic potentials for states with axial quantum numbers $n = 1$ and $n = 2$, marked by solid and dashed lines, respectively, while horizontal lines indicate positions of the lowest subbands bottoms. Comparing curves in Figures 3(a) and (c) with those in Figures 1(a) and (c), one can see that in the presence of the electric field of $15\,kV/cm$ the heights of barriers are reduced in about 100 times, while their widths become essentially narrower, independently on the NT radius size.

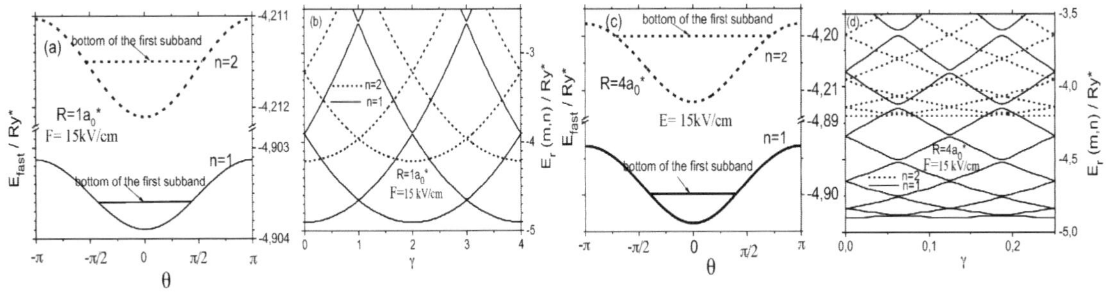

Figure 3. The same as in Figure1 but for the electric field $15kV/cm$.

Consequently, electric field facilitates electron-to-hole tunnelling through barriers, enlarges the widths of subbands, originating their overlapping and the multiple crossovers between curves of the energies dependencies on the magnetic field on the right side of Figure 3. These curves have a behaviour typical for one-particle system in a nanostructure with a ring-like geometry, AB oscillation of the ground state with minima at the points $T = 2n/R^2$; $n = 1,2,3,$, multiple crossovers between curves at the points $T = (2n-1)/R^2$; $n = 1,2,3,$ a successive increase of the amplitudes of oscillation of the excited states and the absence of low-lying states whose energies are independent on the magnetic field. Thus, the external electric field which has a large effect on the Coulomb interaction between the electron and the hole, changing essentially the spectral properties of the magnetoexciton confined in NT.

4. Conclusions

We propose a simple procedure for solving two-particle wave equation describing the exciton confined in a nanotube in the presence of the external magnetic and electric fields applied along the symmetry axis. For zero-electric-field case we found clearly distinct dependencies of lower energy levels on the magnetic field for nanotubes with small and large radius, provided by a transformation of the relative electron-hole arrangement, from the gas-like configuration to the Wigner-like one, with an increase of the nanotube's radius. We find that in the presence of a strong electric field, applied along the symmetry axis, the two-dimensional configuration and the AB oscillations in these states are restored partially or totally, due to the polarization of the exciton in the vertical direction. Our calculations for the dipole moment and the polarizability of the exciton as functions of the electric field show that the ground state configuration of the exciton in nanotubes of large radius undergoes a significant modification under external electric field. Particularly, for the electric field of several kV/cm, the separation between the electron and hole in the axial direction becomes such large that the relative rotation around the axis is "defrosted". Our estimation show that the height of the barriers of the effective potential corresponding to the relative rotation responsible for the AB oscillation of the exciton energies is lowered in 100 times while the electric field is increased up to 15kV/cm. Therefore in this regime the rotation of the electron and the hole around the axis can be considered as almost independent.

Acknowledgements

This work was financed by the «Universidad Industrial de Santander (UIS) » through the «vicerrectoría de Investigación y Extensión (VIE), DIEF de Ciencias, (Cod.5124) » and the «Patrimonio autónomo del Fondo Nacional de financiamiento para la ciencia, la tecnología y la innovación Francisco José De Caldas-Contrato RC –No. 275-2011» and Cod. No.1102-05-16923, subscribed with «Colciencias».

References

[1] A V Chaplik 1995 *JETP Lett* **62**, 900
[2] F Ding, N Akopian, B Li, U Perinetti, A Govorov, F M Peeters, C C BofBufon, C Deneke, Y H Chen, A Rastelli, O G Schmidt, V Zwiller 2010 *Phys Rev B* **82** 075309
[3] J Song and S E Ulloa 2001 *Phys Rev B* **63** 125302
[4] L C Porras, I D Mikhailov 2013 *Physica E* **53** 41
[5] Bin Li and F M Peeters 2011 *Phys Rev B* **83** 115448
[6] K Kubota, P O Vaccaro, N Ohtani, Y Hirose, M Hosoda, T Aida 2002 *Physica E* **13** 313
[7] K Tsumura, Sh Nomura, P Mohan, J Motohisa and T Fukui 2007 *Japan J Appl Phys* **46** L440
[8] F M Peeters and V A Schweigert 1996 *Phys Rev B* **53** 1468
[9] F J Betancur, I D Mikhailov and L E Oliveira 1998 *J Phys D Appl Phys* **31** 3391
[10] B Szafran, F M Peeters, S Bednarek and J Adamowski 2004 *Phys Rev B* **70** 235335

Journal of Physics: Conference Series **743** (2016) 012009 doi:10.1088/1742-6596/743/1/012009

Importance of databases of nucleic acids for bioinformatic analysis focused to genomics

L R Jimenez-Gutierrez[1], C J Barrios-Hernández[2], G R Pedraza-Ferreira[2], L Vera-Cala[2] and F Martinez-Perez[2]
[1] Universidad Autónoma de Sinaloa, Mazatlán, Sinaloa, México.
[2] Universidad Industrial de Santander, Bucaramanga, Colombia.

E-mail: ljimenez@uas.edu.mx , fjmartin@uis.edu.co

Abstract. Recently, bioinformatics has become a new field of science, indispensable in the analysis of millions of nucleic acids sequences, which are currently deposited in international databases (public or private); these databases contain information of genes, RNA, ORF, proteins, intergenic regions, including entire genomes from some species. The analysis of this information requires computer programs; which were renewed in the use of new mathematical methods, and the introduction of the use of artificial intelligence. In addition to the constant creation of supercomputing units trained to withstand the heavy workload of sequence analysis. However, it is still necessary the innovation on platforms that allow genomic analyses, faster and more effectively, with a technological understanding of all biological processes.

1. Introduction

Since the last two decades, bioinformatics has evolved from a field that was not considered one of the branches of science; but now is a key element in the development of basic and applied sciences. On the other hand, its definitions change in function of its execution area; for example, Hogeweg [1] defines bionformatics as the study of processes in biotic systems. Whereas, Tenenbaum [2] shows the translational bioinformatics definition from the American Medical Informatics Association (AMIA): the development of methods of analysis, storage, and interpretation, which optimize the transformation of increasingly genomic and biomedical data, into predictive, and participatory health. Besides, the Dr. Russ Altman from Stanford University, in his 2014 presentation, provided the following definition for bioinformatics: "informatics methods that link biological entities (genes, RNA, proteins, and small molecules) to clinical entities (diseases, symptoms, and drugs) and vice versa".

We consider bioinformatics as a science which final goal is to generate knowledge from biological data, which are analysed from computational systems by mathematic algorithms and/or artificial intelligence. From this perspective, the generation of data with biological importance are placed in codes in a database in computer systems. This is considered the starting point; meaning the success or failure of any bioinformatics analysis.

This principle is critical, because, if the data not accord to the parameters of experts in the field you want to describe and understand, the study model will not explain the reality and accuracy of the system. That is why that the experimental strategy and technological elements that allow data generation are the key elements for achieving results, which when placed in a code database, describing any type of system; specially a biological system.

Content from this work may be used under the terms of the Creative Commons Attribution 3.0 licence. Any further distribution of this work must maintain attribution to the author(s) and the title of the work, journal citation and DOI.
Published under licence by IOP Publishing Ltd

2. Biological aspects and experimental developments which has allowed obtain the nucleic acids and protein sequences; to form the basis of current data

Gregor Johann Mendel could be considered the originator and systematizer of the biological data obtained related to inheritance. He assigned a numerical value to his results in plants hybridization to apply their mathematical models and explain their results [3]. Subsequently, a number of discoveries were instrumental in the ongoing development of bioinformatics of nucleic acids. The first, corresponds to the purification of deoxyribonucleic acid by Friedrich Miescher with 5 papers among 1892-1897 [4], he discover the nucleus as the DNA carrier. Phoebus Levine subsequently discovered that the nucleic acids are composed of a sugar: ribose or deoxyribose, and 5 types of bases: adenine, thymine, guanine, cytosine and uracil [5]. Edwin Chargaff [6] describes that they are associated in proportions 1:1 adenine - thymine, adenine - uracil and guanine - cytosine. After that, Avery [7]; Hershey and Chase [8], demonstrated that nucleic acids are molecules containing genetic information in bacterial and virus, respectively, and Watson and Crick [9] proposed the double helix model of DNA.

With all these biochemical and chemical background, in 1977 were published two methods to establish the nucleotide sequence of a DNA segment: one proposed by Allan M. Maxam and Walter Gilbert [10]; which was based on the hydrolysis of DNA. Meanwhile, the second proposed by Frederick Sanger et al [11] with a nucleic acid sequencing performed by incorporating radioactive nucleotides into DNA chains and then separated by electrophoresis. The relative simplicity of implementation of cither sequencing method allows obtaining nucleic acid segments, quickly and easily, and with thus, the implementation of codes for its analysis. But with the growing number of nucleotide sequences, it was necessary to provide a solution to a seemingly inconsequential problem, but fundamental; the development of a communication system of the nucleotide sequences for study of any research group.

3. Development of computing and informatics applied to building databases

Bioinformatics is a multidisciplinary domain with a traditional and large use of high performance computing capabilities. In fact, the idea to improve knowledge with the innovative application of computer science and computer resources allows the possibility of treatment of data in life sciences. High Throughput data, sequencing and data visualization are today bio-informatics technologies linked with laboratory experiences and theory.

Recently, EEUU presidents Bush and Obama, in 2004 and 2010 respectively, proposed improvement of computational resources to understand human biology addressed to personalized healthcare. Then, projects as human genome [12], human brain [13], Parkinson Modelling Project, human heart model, among others, are developed faster in EEUU, with a worldwide participation; mainly of the European Union and China. These projects demand an important use of high performance computing centres to use de genomic database results.

If we observe the Top500.org site [14], which contains the 500 supercomputers with the most important performance of the world, it is possible to distinguish that the 48% of these supercomputer platforms are used for research, of which 50% is related to life sciences. Bioinformatics with corresponding databases for any field puts enormous challenges to computer sciences, this challenge implies five important subjects: Complexity of the System, Size of the Data, Processing and Data Flow, Heterogeneity of the Data and Computing Power. The complexity of the systems must be treated with the corresponding algorithms and implementation mechanisms. study is necessary to propose solutions, and of course, a posterior implementation in a computing algorithm.

The size of the data has two points of views: technology for storage and security requirements, and mathematical (related to complexity). In the first point of view, the capacity and safety of the technological solutions are important to store the data. The second, same than the complexity, algorithms and mathematic process are important, for example, using computational techniques; such as big data or deep learning for large scale databases.

Processing and Data Flow are treated using software engineering techniques, and of course, the processing is transversal to the algorithm implementation and coherence of the results. At the same time, it is important to observe the data heterogeneity. This involving the nature of the algorithms, the use of non-structured formalization of the data and its implementation, and at same time, technological computer architecture. Finally, the computing power improves the performance, and of course, it allows to obtain results in predictable short times.

For genomics, databases involve sequences, maps, annotations and chromosomes. Different available resources have different sizes and formats, then for the analysis and computer treatment is important to observe the format and provide "simple" standard results. The large size of the datasets and the heterogeneity requires specific techniques to make an implementation of algorithms and codes. On the other hand, the visualization of the large scale data and results is important; however, in this paper this subject is not shown.

The particularity of the datasets and the high-granularity of the data and related tasks to genomic studies allow implementations in heterogeneous computer platforms, based in GPUs (Graphical Processing Units); for example, the GUANE platform (GpUs AdvaNced computing Environment), a 128 Fermi TESLA GPUs supercomputer based in the High Performance and Scientific Computing of the Universidad Industrial de Santander, Colombia. The implementation of the algorithms is developed in NVIDIA(r) CUDA(r) [15]; a programing model proposed by NVIDIA to implement mechanisms of exploitation and implementation of granularity over GPUs. This architecture allows to treat complexity, large scale and heterogeneity of data.

4. Nucleic acids and genomic database

Due to the high number of reports of sequences of nucleic acids and proteins generated by the old methodologies and new technological advances, in 1979, 35 scientists meeting at Rockefeller University, they concluded the urgency of establishing a nucleic acids International database, which could correlate, as much as possible, the biological information; concluding with the creation of the GenBank. After this meeting the international scientific community continued with similar events [16, 17]. However, with the increase of nucleotide sequences, in 1987 was created The International Nucleotide Sequence Database Collaboration (INSDC), which currently consists of the DNA Data Bank of Japan (DDBJ) at the National Institute for Genetics in Mishima, Japan; the European Nucleotide Archive at the European Bioinformatics Institute (EMBL) in Cambridge, UK; and GenBank at the National Centre for Biotechnology Information (NCBI) in Bethesda, MD, USA [18, 19].

Since its creation, the INSDC guideline and policy is keeping nucleotide sequences obtained by experimental methods in public databases, and allow unrestricted access to any person or institution. In this sense, in 2015 this database was constituted by 2.400 trillion bases [19]. This amount will increase in the next years thanks to new genome sequencing technologies developed by Pacific Biosciences, Illumina, Roche or Ion Torrent [20, 21]; besides to the new Computer platforms and computational programs that will allow the completion of projects such as personalized genomic medicine [21], food, and basic research projects.

But even with this genomic revolution, that will increase the nucleotide databases, the challenge remains finding genomes regions whose composition and base order has not allowed their obtaining. For this purpose, the scientific community is developing technologies to increase capacity and fidelity of polymerization performs by DNA polymerases and reduce assembly times of genomes and their products. The solution to these and other technological processes will allow increase the genomic databases, but the biggest challenge is the proper use and application of the nucleotide sequences, which will be built for humanity benefit.

References

[1] Hogeweg P 2011 The roots of bioinformatics in theoretical biology *PLo. S. Comput. Biol.* **7(3)** e1002021

[2] Tenenbaum J D 2016 Translational Bioinformatics: Past, present, and future *Genomics Proteomics Bioinformatics* **14(1)** 31-41

[3] Mendel G 1866 Versuche über Pflanzen-Hybriden *Verhandlungen dernaturfoschung Vereins* **4** 3–47

[4] Dahm R 2008 Discovering DNA: Friedrich Miescher and the early years of nucleic acid research *Hum. Genet.* **122(6)** 565-581

[5] Levene P A 1903 On the chemistry of the chromatin substance of the nerve cell *J. Med. Res.* **10(2)** 204-211

[6] Chargaff E 1951 Some recent studies on the composition and structure of nucleic acids *J. Cell. Physiol. Suppl.* **38(1.1)** 41-59

[7] Avery O T Macleod C M McCarty M 1944 Studies on the chemical nature of the substance inducing transformation of Pneumococcal types: induction of transformation by a desoxyribonucleic acid fraction isolated from Pneumococcus type III *J. Exp. Med.* **1(79-2)** 137-58

[8] Hershey A D Chase M 1952 Independent functions of viral protein and nucleic acid in growth of bacteriophage *J. Gen. Physiol.* **36(1)** 39-56

[9] Watson J D Crick F H 1953 Molecular structure of nucleic acids; a structure for deoxyribose nucleic acid *Nature.* **171(4356)** 737-738

[10] Maxam A M Gilbert W 1977 A new method for sequencing DNA *Proc. Natl. Acad. Sci. USA* **74(2)** 560-564

[11] Sanger F Nicklen S Coulson A R 1977 DNA sequencing with chain-terminating inhibitors *Proc. Natl. Acad. Sci. USA.* **74(12)** 5463-5467

[12] National Human Genome Research Institute 2016 https://www.genome.gov/

[13] The Human Brain Project (THBP) 2016 https://www.humanbrainproject.eu/

[14] Top 500 List 2016 http://www.top500.org

[15] NVIDIA, CUDA 2016 http://www.nvidia.com/object/cuda_home_new.html

[16] Smith T F 1990 The history of the genetic sequence databases *Genomics.* **6(4)** 701-707

[17] Strasser B J 2008 Genetics. GenBank-Natural history in the 21st Century? *Science* **322(5901)** 537-53

[18] Cochrane G Karsch-Mizrachi I Nakamura Y 2011 International Nucleotide Sequence Database Collaboration in The International Nucleotide Sequence Database Collaboration *Nucleic. Acids. Res.* **39** D15-D18

[19] Cochrane G Karsch-Mizrachi I Takagi T 2016 Sequence Database Collaboration in The International Nucleotide Sequence Database Collaboration *Nucleic. Acids. Res.* **4(44)** D48-50

[20] Liu L Li Y Li S Hu N He Y Pong R Lin D Lu L Law M 2012 Comparison of next-generation sequencing systems *J. Biomed. Biotechnol.* **2012** 251364

[21] Quail M A Smith M Coupland P Otto T D Harris S R Connor T R Bertoni A Swerdlow H P Gu Y 2012 A tale of three next generation sequencing platforms: comparison of ion torrent, pacific biosciences and illumina MiSeq sequencers *BMC Genomics.* **13** 341

[22] Linderman M D Nielsen D E Green R C 2016 Personal genome sequencing in ostensibly healthy individuals and the peopleSeq consortium *J. Pers. Med.* **6(2)** E14

Journal of Physics: Conference Series **743** (2016) 012010 doi:10.1088/1742-6596/743/1/012010

A quantum chemical study for exploring the inhibitory effect of nitrogen containing species on the adsorption of polynuclear aromatic hydrocarbons over a Brønsted acid site

C M Celis-Cornejo[1], M M Garnica Mantilla[1], V G Baldovino-Medrano[1] and G E Ramírez-Caballero[1]
[1] Universidad Industrial de Santander, Bucaramanga, Colombia.

E-mail: ransereg@hotmail.com ; gusramca@uis.edu.co

Abstract. The analysis of the inhibitory effect of nitrogenated compounds on the hydroprocessing and hydropurification of oil derived fuels is important to produce cleaner fuels. In this work, density functional theory calculations were performed to investigate the effect of the nitrogen containing molecules on the adsorption of Polynuclear Aromatic Hydrocarbons (PAHs). Mordenite was chosen as a zeolitic structure for simulating a Brønsted acid site. The character of the acid site was confirmed by both a vibrational frequency calculation and a Bader charge analysis. From the adsorption calculations, it was found that the adsorption energy of PAHs increases with the number of aromatic rings in the structure. Also, the nitrogen containing species possibly inhibit more extensively two and three rings PAHs because of their lower adsorption energies. Finally, it was observed that the nitrogen species tend to drag the proton from the mordenite acid site. This explains the inhibitory effect in the adsorption of PAHs and contributes to understanding the dynamics of hydrocarbon hydroprocessing in refineries.

1. Introduction

Hydrocracking is key for transforming heavy oil feedstocks into desired liquid fractions such as gasolines and diesel. The heavier the fraction, the higher the concentration of heteroatoms [1]. Hydrocracking is usually preceded by hydrotreating in which sulfur, nitrogen and metal heteroatoms are removed [2]. After hydrotreating some refractory heteroatom species remain, nonetheless.

A previous study on the characterization of a Vacuum Gas Oil (VGO) and of its hydrotreated product (HDT-VGO) by Electrospray Ionization Fourier Transform Ion Cyclotron Mass Spectrometry (ESI FT-ICR-MS) showed that the most abundant nitrogen containing species in this VGO have a N_1 stoichiometry containing one nitrogen atom in their structure [3]. Furthermore, this study also showed that basic nitrogen species mostly comprise pyridine type structures such as quinolines and benzoquinolines in addition to non-basic nitrogen species which consist mainly on pyrrolic type structures such as carbazoles and benzocarbazoles [4]. Between these two nitrogenated families, the basic species are easier to remove during hydrotreatment as compared to the non-basic ones [3].

The effect of basic nitrogen molecules on the catalytic performance of hydrotreating materials has been copiously studied. Dufresne *et al.* [5] reported that it is necessary to increase 100°C the reactor temperature in order to level off the conversion in the reactor at 6.1kPa when increasing ammonia partial pressure. Molecules of this nature interact strongly with both Brønsted and Lewis acid sites. Korre *et al.* [6] found that the adsorption constants for ammonia on catalysts acid sites are two orders

Content from this work may be used under the terms of the Creative Commons Attribution 3.0 licence. Any further distribution of this work must maintain attribution to the author(s) and the title of the work, journal citation and DOI.
Published under licence by IOP Publishing Ltd

of magnitude higher than those calculated for the hydrocracking of Polyaromatic Aromatic Hydrocrbons (PAHs). Sau *et al.* [7] established that concentrations of pyridine as low as 10ppm have a detrimental effect on hydrocracking [7]. Concerning the effect of non-basic nitrogen species on hydrocracking catalysts, the topic remains basically unexplored under the common assumption that only the effects of basic nitrogen species on hydrocracking are relevant [8]. However, Kobayashi *et al.* [9] reported a strong effect of non-basic carbazole over the activity and selectivity of hydrocracking catalysts even when the concentration of this molecule was not higher than 20ppm.

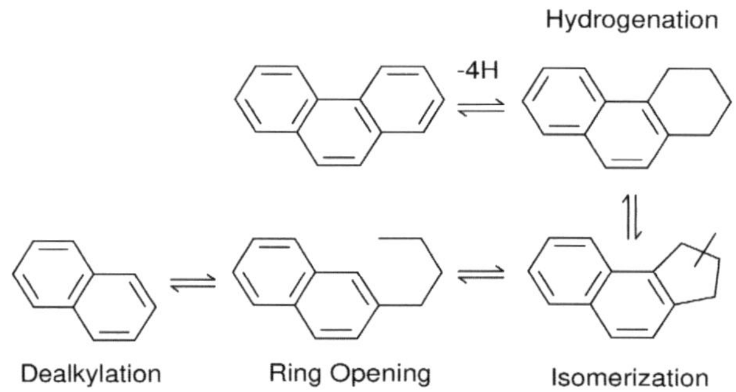

Figure 1. Commonly reported hydroconversion reaction route of phenanthrene into naphthalene.

Phenanthrene and naphthalene are model molecules for heavy feedstocks. The hydrocracking of phenanthrene has thus been studied over Pt [10], FeW [11], NiW [6] and NiMo [12] based catalysts as well as over catalytic supports such as □–Al_2O_3 [12], HY and HZMS-5 zeolites [10], MCM-41 [10], and Al-SBA-15 [11]. In particular, studies over NiMo/HY catalysts have shown that phenanthrene hydrocracking is preceded by partial hydrogenation to intermediates such as: 1,2,3,4-Tetrahydrophenantrhene, 9,10-Dihydrophenantrhene and 1,2,3,4,5,6,7,8-Octahydrophenantrhene over the hydrogenating metal sites of the catalyst [6]. It is only after these reactions that isomerization, ring opening and dealkylation steps proceed over the supports Lewis and Brønsted acid sites. Figure 1 shows the hydrocoversion route for the 1,2,3,4-Tetrahydrophenanthrene into naphthalene as under the above considerations.

Up to this point, the above discussion points out the importance of understanding what would be the inhibitory effect of non-basic nitrogenated molecules on the adsorption and reactivity of PAHs such as phenanthrene. Such understanding is important for perfecting kinetic modeling of the hydroconversion of heavy feedstocks. Consequently, we decided to make a first approach to the problem by performing Density Functional Theory (DFT) calculations for the adsorption of partial hydrogenated polynuclear hydrocarbons and of basic quinoline and non-basic carbazole over a model Brønsted acid site within a mordenite type zeolitic structure which is commonly used in hydrocracking.

2. Molecular simulation

In this section, the procedure for modeling the catalyst surface and the adsorption of the studied molecules are presented.

2.1. Mordenite structure

Mordenite has a unit cell containing 144 atoms. In comparison, faujasite and ZSM-5 [10] zeolites, also commonly used in hydrotreatment, contain 600 and 280 atoms, respectively. Therefore, calculations on mordenite have lower computational cost. Natural mordenite has the chemical formula: $Na_8Al_8Si_{40}O_{96}*24H_2O$, an orthorhombic morphology, a Cmcm space group, and crystallography unit cell parameters a=18.052-18.168Å, b=20.404-20.527Å and c=7.501-7.537Å. Mordenite is a one-dimensional pore zeolite with ring sizes of 12, 8, 5 and 4-membered rings. The bulk mordenite CIF file was taken from the Database of Zeolite Structures [13]. Initial unit cell parameters were a=18.256Å, b=20.534Å, and c=7.542Å. For the sake of convenience, the adsorption of the selected compounds was simulated over the 12-membered ring which has a cross section between 12.16Å and 11.25Å. Figure 2 shows the aforementioned physicochemical properties. After estimating the simulation parameters, a surface was created adding a vacuum space of 15Å over the [001] plane in order to expose the 12-membered ring. Terminal oxygen atoms were completed with hydrogen atoms. Following this procedure, we obtained a structure with 193 atoms.

2.2. Method
Calculations were carried out with the Vienna Ab-Initio Simulation Package (VASP) considering periodic boundary conditions. The Perdew–Burke–Ernzerhof exchange correlation functional (PBE) was considered in all cases. All of the simulation parameters were first estimated for the bulk mordenite system with smearing 0 and sigma parameter of 0.05. The energy cut off was also estimated initially at 1.3 times of the pseudopotential ENMAX value. A 400eV was chosen after a convergence study. A Γ-point Brillouin-zone sampling was considered due to the large size of the unit cell.

2.3. Acid site simulation
Commercial mordenite usually has Si/Al ratios ranging from 6 to 10 while some commercial Y zeolites have a Si/Al ratio from 5 to 37. For adding a Brønsted acid site to the zeolitic structure, the isomorphic substitution of one silicon atom by one aluminum atom was made in a structure with a Si/Al ratio of 47 (Figure 3). Such low Si/Al ratio is considered herein as representative of a very dealuminated acid support possessing an intermediate acidity such as those commonly used for hydrocracking. A vibrational frequency calculation was carried out to confirm the acid character of the simulated site. A Bader Topological Analysis was also performed using the algorithm developed by Henkelman and coworkers [14] for obtaining the accumulated charge of the atoms in the system and in order to estimate the electronic charge of the modeled acid site.

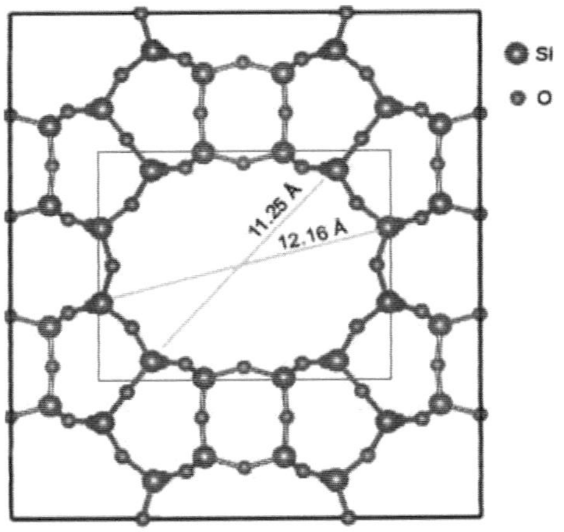

Figure 2. Bulk 144 atom unit cell of mordenite from a top view.

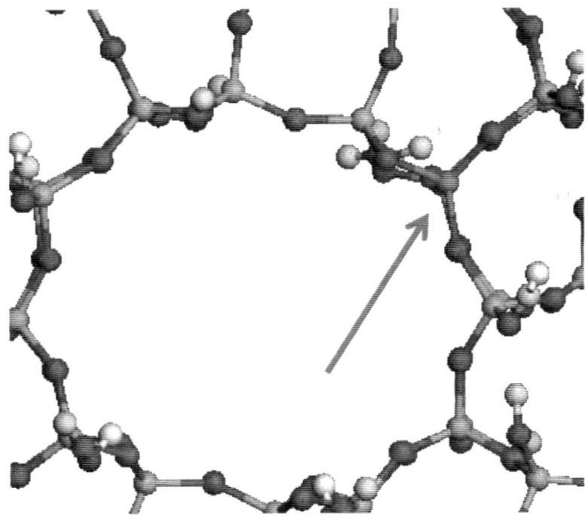

Figure 3. Catalytic surface top view of the [001] plane, aluminum in purple.

2.4. Adsorption of PAHs and nitrogenates
As discussed before, partial hydrogenation of the PAHs over the metallic active sites of the catalysts is a prerequisite for isomerization, ring opening, and dealkylation reactions. Therefore, only the adsorption of partially hydrogenated hydrocarbons such as 1,2,3,4-tetrahydronaphthalene, 1,2,3,4-tetrahydrophenanthrene, and 1,2,3,4-tetrahydrocrysene, 1,2,3,4-tetrahydrocarbazole, and quinoline and carbazole, over the model acid site were considered.

Each adsorbate was firstly optimized in a 25Å vacuum cube cell for avoiding interactions between the periodic images hence obtaining the energy of the isolated molecules (E_m). Subsequently, the adsorption energy (ΔH_{ads}) was calculated with the following Equation 1:

$$\Delta H_{ads} = E_{m-zeo} - (E_m + E_{zeo}) \quad (1)$$

Where, E_{zeo} and E_{m-zeo} represent the mordenite catalytic surface and the mordenite-adsorbate complex energies, respectively.

Partially hydrogenated PAHs were adsorbed by the C' carbon (see, Figure 4(a)), carbazol, its partially hydrogenated product, and quinoline were adsorbed by the N' nitrogen (see Figure 4(b) and 4(c)). The choice of the latter was based on the fact that nitrogen has in these molecules a basic electron pair. Initial configurations were set at 1.1Å for each interaction considered. All of the calculations were relaxed without restrictions until all adsorbates reached an equilibrium distance from the acid site.

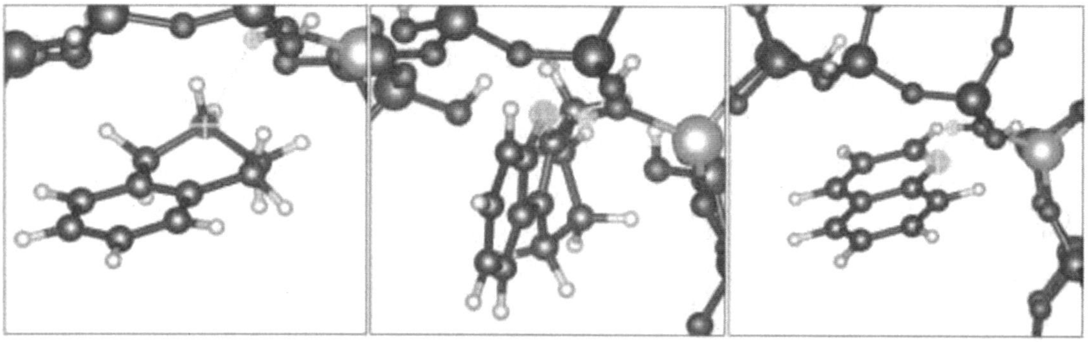

(a) (b) (c)

Figure 4. Adsorption orientations: (a) 1,2,3,4-tetrahydronaphthalene, (b) 1,2,3,4-tetrahydrocarbazole, and (c) quinoline.

3. Results and discussion

3.1. *Catalyst surface and acid site*

The parameters of the unit cell obtained after mordenite optimization were a=18.094Å, b=20.516Å, and c=7.524Å. From the stretching frequency calculations for the OH adjacent to the aluminum atom, a frequency of 3603.9cm^{-1} was obtained. This value very close to that reported by IR studies of Brønsted acid sites (3608cm^{-1}) [15]. The Bader charge analysis yielded a total charge of 896; this value corresponds to the external electrons considered by the Effective Core Potential. The accumulated charge for the hydrogen attached to the oxygen adjacent to the aluminum atom is about +0.75. This value is higher than that for terminal silanol groups that average +0.64. These results confirm the acid character of the simulated Brønsted acid site.

3.2. *Adsorption of PAHs and nitrogen containing molecules*

Table 1 shows that the adsorption energy for carbazole is higher than that of its partially hydrogenated product. Similar experimental [6] and theoretical [16] results have been reported. Now, the adsorption energy calculated for 1,2,3,4-tetrahydrocarbazole is higher than that for 1,2,3,4-tetrahydrophenanthrene as well as for 1,2,3,4-tetrahydronaphthalene. Hence suggesting a possible inhibitory effect of this nitrogen-containing molecule over the adsorption of species with three or two rings. On the other hand, 1,2,3,4-tetrahydrocrysene had a higher adsorption energy whereby it could be less inhibited by the presence of non-basic nitrogen containing species. The calculated trend for the adsorption energies for PAHs increases as long as aromatic rings are added to the structure, therefore four ring PAHs displayed the highest adsorption energy. Thermochemical experiments for the hydrocracking of phenanthrene evidence that the equilibrium constant for 1,2,3,4-tetrahydrophenanthrene is higher than the one calculated for 1,2,3,4-tetrahydronaphthalene [6]. Therefore, our results agree well reported experiments. Finally, the quinoline adsorption energy was highest as in agreement with a strong interaction between its basic nitrogen heteroatom and the model Brønsted acid site.

Table 1. Adsorption energies calculated from Equation 1.

Molecule	ΔH_{ads} (kJ/mol)
Carbazole	-486.77
1,2,3,4-Tetrahydrocarbazole	-481.53
Quinoline	-570.63
1,2,3,4-Tetrahydronaphthalene	-382.93
1,2,3,4-Tetrahydrophenanthrene	-414.93
1,2,3,4-Tetrahydrocrysene	-457.29

Equilibrium distances calculated for the interactive atoms after relaxation are shown in Table 2. For PAHs, equilibrium distances were reached at 2.23, 2.15 and 2.07Å for 1,2,3,4-tetrahydronaphthalene, 1,2,3,4-tetrahydrophenanthrene, and 1,2,3,4-tetrahydrocrysene, respectively. Calculations show the displacement of all of the molecular structure due to a repulsive force until reaching an optimum distance. The herein obtained equilibrium distances are in good agreement with other theoretical works using the General Gradient Approximation (GGA) [17]. García-Serrano *et al.* [17] calculated distances of 2.16 and 2.20Å for the carbon-proton interactions. It should be noticed that the proton remains at the same initial position. For both non-basic nitrogenates, similar equilibrium distances were estimated but with a displacement of the proton. The latter is clearly attracted by the nitrogen heteroatom. Equilibrium distances for quinoline also show such an effect but with the proton closer to the nitrogen heteroatom and at a distance of 1.06Å which is almost the same as that for a N-H bond. This suggests that the inhibition occurs not only because the distances are shorter between nitrogen (N') and the proton (H+) but also because nitrogen draws the proton out of the mordenite structure. These trends agree with the known stronger inhibitory effect of basic nitrogen during hydrocraking reactions.

Table 2. Equilibrium distances for the interactive atoms after geometry optimization for all of the adsorbates.

Molecule	Interaction	Distance (Å)
Carbazole	N'---H+	1.13
	H+---O	1.45
1,2,3,4-Tetrahydrocarbazole	N'---H+	1.12
	H+---O	1.45
Quinoline	N'---H+	1.06
	H+---O	1.51
1,2,3,4-Tetrahydronaphthalene	C'---H+	2.20
	H+---O	0.99
1,2,3,4-Tetrahydrophenanthrene	C'---H+	2.14
	H+---O	0.99
1,2,3,4-Tetrahydrocrysene	C'---H+	2.09
	H+---O	0.99

Finally, a Bader charge analysis of the complex quinoline-mordenite at the equilibrium distance shows that the charge of the proton is +0.51 while that of the neighboring oxygen shifts from -1.61 to -1.52. In the meantime, the charge for the nitrogen heteroatom shifts from -1.15 to -1.26. Consequently, a strong interaction between the proton moieties and the basic nitrogen of quinoline is well evidenced.

4. Conclusions

This work is a first theoretical approach to revealing the effect of non-basic nitrogenates on the hydrocracking of heavy oil fractions. Particularly, adsorption of non-basic and basic nitrogenated molecules such as carbazole, 1,2,3,4-tetrahydrocarbazole and quinolone, over a single Brønsted acid site, located at the surface of a mordenite structure was simulated. Calculations showed that the increase in the number of aromatic rings of the molecules leads to higher adsorption energies (ΔH_{ads}). Nonetheless, the adsorption energy for the non-basic nitrogen species shows a possible inhibitory effect due to competition for the acid site with PAHs mainly with 2 and 3 rings. The shift observed in the position of the proton locating this closer to the nitrogenated adsorbate than to the mordenite evidences the inhibitory effect of nitrogenated compounds for the adsorption of PAHs in Brønsted acid sites.

Acknowledgements

Authors thank to the VIE-UIS for funding. C M Celis-Cornejo thanks the Instituto Colombiano del Petróleo (ICP) from Ecopetrol S.A. and Colciencias for funding his PhD. The gentle collaboration of the SC3 work team from the GUANE platform at UIS is also acknowledged.

References

[1] Motaghi M, Ulrich B and Subramanian A 2011 *Hydrocarb. Process* **90** 37-43
[2] Paul R. Robinson, Geoffrey E. Dolbear 2006 *Practical advances in petroleum processing. Springer* vol 1 ed Chang Samuel Hsu, Paul Robinson (New York: Springer-Verlag) pp 177-218
[3] Kekäläinen T, Pakarinen J M H, Wickström K and Vainiotalo P 2009 *Energ. Fuel* **23** 6055-6061
[4] Jokuty P L and Gray M R 1991 *Energ. Fuel* **5** 791-795
[5] Dufresne P, Quesada A and Mignard S 1989 *Stud. Surf. Sci. Catal* **53** 301-315
[6] Korre S C, Klein M T and Quann R J 1997 *Ind. Eng. Chem. Res* **36** 2041-2050
[7] Sau M, Basak K, Manna U, Santra M and Verma R P 2005 *Catal. Today* **109** 112-119
[8] Prins R 2001 *Advances in Catalysis: Academic Press* **46** 399-464
[9] Kobayashi M, Togawa S and Ishida K 2007 *J. Jpn. Pet.* **50** 44-52
[10] Benazzi E, Leite L, Marchal-George N, Toulhoat H and Raybaud P 2003 *J. Catal* **217** 376-387
[11] Restrepo-Garcia J R, Baldovino-Medrano V G and Giraldo S A 2016 *Appl. Catal. A* **510** 98-109
[12] Lemberton J-L and Guisnet M 1984 *Appl. Catal.* **13** 181-92
[13] Baerlocher C and McCusker L B 2007 *Database of Zeolite Structures* (http://www.iza-structure.org/databases/)
[14] Tang W, Sanville E and Henkelman G 2009 *J. Phys. Condens. Matter* **21** 084204
[15] Bucko T, Benco L, Demuth T and Hafner J 2002 *J. Chem. Phys.* **117** 7295-305
[16] Song T, Zhang Z, Chen J, Ring Z, Yang H and Zheng Y 2006 *Energ. Fuel* **20** 2344-2349
[17] García-Serrano L A, Flores-Sandoval C A and Zaragoza I P 2003 *J. Mol. Catal. A: Chem.* **200** 205-212

Journal of Physics: Conference Series **743** (2016) 012011 doi:10.1088/1742-6596/743/1/012011

Molecular catchers for pharmacologically active substances in wastewaters, a theoretical study

P J Salazar Valencia[1], S T Pérez Merchancano[1], H Paredes[2] and L E Bolívar Marinez[1]
[1] Universidad del Cauca, Popayán, Colombia.
[2] Universidad Industrial de Santander, Bucaramanga, Colombia.

E-mails: pjsalazar@unicauca.edu.co

Abstract. A basic and pressing need in the treatment of residual waste waters for urban and rural centers is the removal of pharmacological active residues from them, these resides are originated in a wide array of domestic, agricultural and industrial sources and can´t be removed in the residual waters treatment plants by conventional methods, the result is the incorporation of them into the ecosystem altering the physiology and behavior of living organisms. Among the most active pharmacological substances found in very high concentration in residual waters is paracetamol, an analgesic of very wide excessive use due to its ease of access and low cost [1]. No pharmacological substance is entirely absorbed by the human organism and therefore a wide family of molecular residues is excreted by the urinary tract. In this work we have used the AM1 (Austin Model 1), PM3 (Parametric Method 3) and ZINDO/CI semiempirical methods, from the NDO (Neglect Differential Overlap) family [2] to study and observe the structural, electronic and optical characteristics of paracetamol while immersed in different basic and acidic aqueous environments, either alone or interacting with lignosulphonates. We have previously found that lignosulphonates, a lignin derivatives of wide industrial applications, can be engineered as a binding and flocculant agent and acts as molecular catchers therefore showing the potential to be used as a mean to filter and eliminate molecular residues from the residual waters [3].

1. Introduction

An important problem in the water treatment for human consumption is the increasingly presence of significant amounts of pharmacologically molecular active contaminants or SFA substances, which are incorporated into the water cycle after being excreted by living beings through sewage water from domestic, industrial, and clinical origin. Among these SFA one of the most abundant corresponds to prescription free painkillers, which are widely used due to their low cost and wide availability, one of them is paracetamol, commercially known as acetaminophen.

Acetaminophen is a drug with analgesic properties, which acts by inhibiting the synthesis of prostaglandins, cellular mediators that are responsible for the onset of pain, has good antipyretic (fever control) but no substantial anti-inflammatory properties, it also has a wide spectrum of commercial presentations.

This compound comes from N-acetyl-para-aminophenol and acetyl-para-aminophenol and in Figure 1 we can see its molecular structure [4].

Content from this work may be used under the terms of the Creative Commons Attribution 3.0 licence. Any further distribution of this work must maintain attribution to the author(s) and the title of the work, journal citation and DOI.
Published under licence by IOP Publishing Ltd

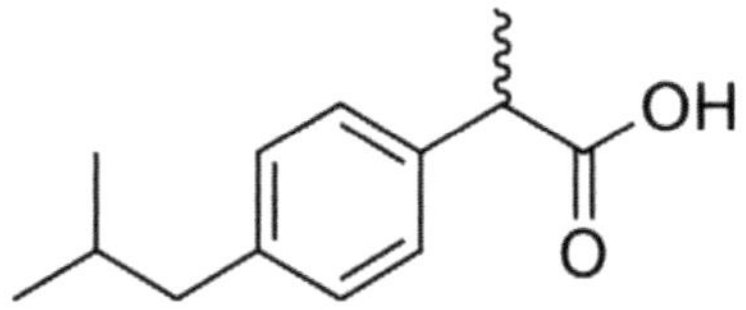

Figure 1. Paracetamol (acetaminophen) Structure.

Paracetamol is mainly metabolized in the liver, resulting in some cases in non-toxic products by means of glucuronidation (glucuronic acid addition) and sulfating, and in other cases generating toxic products such as NAPQI or N-acetyl-p-benzoquinone imine.

Lignosulfonates (LS), also called lignosulfonates, lignin sulfonates or lignin sulfites are essentially lignins in sulfonated forms, obtained from "sulfite liquors" which in turn are a residue of the pulp extraction process in the wood pulping industry. They can be used as additive agents, dispersants, binders, fluidificants, agglutinative, etc. Since they can be manufactured relatively easily and quickly, and that its molecular size can be manipulated to obtain fragments of very low molecular weight, they are used as transport agents in the food, cosmetics, pharmaceutical industry and in the development of drugs and molecular elements for health problems treatment [5].

The LS extraction process is known as delignification and acts by breaking the ester bonds (C-O-C) that connect the lignin constituent elements, then the electrophilic carbonations produced during the breaking of ester bonds react with bisulfite ions to produce the sulfonated at the same time incorporating functional groups to the molecular structure. Among the incorporated ions are for example, calcium, sodium, magnesium and ammonium, this process can be schematized briefly in Figure 2 [6].

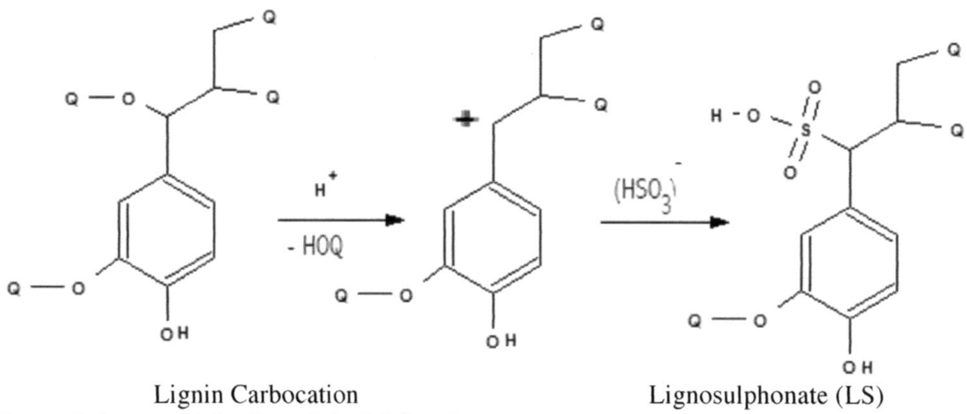

Figure 2. Structural Outline of the LS forming process.

In this process, the woody adhering tissue components are eliminated, resulting in a class of polyelectrolytes' polymers with high acidity due to the presence in their molecular structure groups of highly dissociated sulfonic acid, and unlike lignin when it is present in the cell walls this are water soluble, this family of molecules is collectively known with the name of lignins.

It has been empirically observed, that LS possess capabilities that make them behave in a very effectively way as free radical hunters and antioxidants [7]. The properties of the LS depend, as in any molecular system, from its structure and the characteristics of the individual components, sharing with lignin the basic precursors form and bond mechanisms [8]. By observing the molecular structure we can see a series of small aggregates of 2 or 3 precursor units linked through Carbon Oxygen bonds, among this there is a high proportion of the β-O-4 linking mechanism, that connects the aromatic parts with the carbon chains, in this chains other substituent groups adhere, which in the case of the LS are sulfonates (functional groups containing SO_3 ions); other functional groups adhere to either sulfonates, available positions in the carbon chain or simply appear in molecular or ionic form near the structure.

The macromolecule grows when it polymerizes randomly until an amorphous and three dimensional biopolymer is formed, which, like lignin, is difficult to study because of their size and to the randomness of its formation. Previous studies have shown that in the case of lignins in general, the properties of the basic units or precursors can easily be extrapolated, and be used to describe certain characteristics and behavior of the macrostructure [9]. In this work it was established that the lignosulphonates or sulfonated lignin, residues from the paper pulp extraction process that can be used as binders and flocculating agents of wide industrial use, adhere and encapsulate to the analgesics, paracetamol in particular, fulfilling the function of filter and frame grabbers of molecular pharmacologically active waste in sewage water, for it we use AM1 (Austin Model 1) semi-empirical methods, PM3 (Parametric method 3) and ZINDO / CI, (HyperChem 5 from the NDO family (Neglect Differential Overlap)) that characterizes the compounds in structural, electronic and optical way in different aqueous environment.

2. Methodology

The geometric structure of the compounds with which we worked are known from the literature, the paracetamol molecule is combined with small fragments of up to three (3) LS basic units containing functional groups of Ammonium, Sulfur, Calcium and Sodium in aqueous environments, acidic and basics to study the interaction among molecular systems through specific characteristics such as atomic distances, heat or enthalpy of formation, simulations of electrostatic potentials, etc. The studies were conducted using the molecular modeling program Hyperchem 5 [10] using semiempirical methods AM1 and PM3 from the NDO (Neglect of Differential Overlap) family for calculating the structural and electronic properties and in the case of the optical properties it was used the semiempirical method ZINDO/CI.). A convergence gradient criterion type RMS with 10^{-3} limit was used in all calculations obtained. Electronic and optical properties were calculated in Single Point mode with 6-occupied orbitals and 6 unemployed criterion seeking energy stability of the structure. Paracetamol was structurally characterized using the mentioned methods in several environments (vacuum, acidic and basic aqueous). Then we proceeded to combine the paracetamol molecule with the LS and determine again the molecular systems characteristics consequently formed in each of the environments. For modeling of different environments in which the structures are embedded a periodic box was designed, with 50 water molecules in random positions to make up the environment, and then in it, it submerges the paracetamol and the lignosulphonate considered.

3. Results and discussion

Table 1 shows the values of the heat of formation enthalpy of paracetamol obtained by geometries optimizations using the three main methods of NDDO approach, the best value is provided by the PM3 method, with a difference of about 10kcal/mol respect to the AM1 and MNDO methods. The heat of formation enthalpy indicates how much energy is needed to create the structure, and as more negative the value is, greater will be the probability that the structure exists and be stable, these results are presented in their neutral state and in vacuum.

Table 1. Heat of Formation in kcal/mol for paracetamol in neutral and empty state (vacuum).

Method	Heat of Formation
MDNO	-58.01
AM1	-56.97
PM3	-67.12

Since PM3 offers the best results it was decided to use as a basis for the development of the other optimizations geometries in this work. A representation of the geometrical structure obtained from these calculations can be seen in Figure 3.

The graphical representations of the geometries obtained for the studied molecular systems in this paper are two-dimensional projections of three-dimensional structures, for example the geometry obtained shows a seemingly flat molecule of about 6.5Å in length by 2.42Å wide, with a small side chain leaving the plane with a twist angle of -27° and a molecular mass of 151.15Da (atomic mass units). Figure 3 shows this configuration.

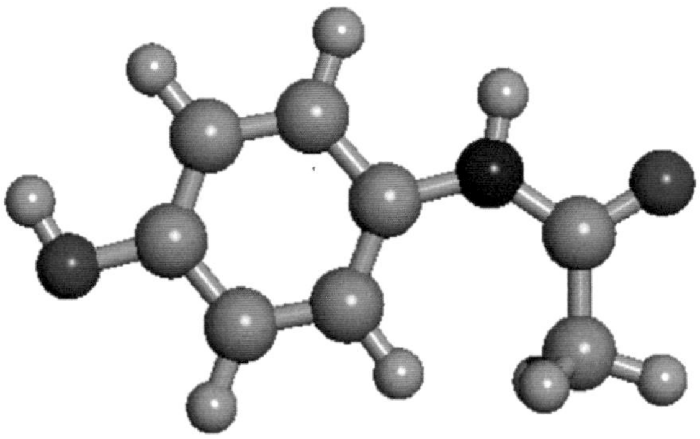

Figure 3. Molecular structure of Paracetamol obtained by optimizing the geometry with the semiempirical method PM3.

To determine the preferential charge state of the structure Heat of Formation is determined for each one of its charging conditions in vacuum and water. The results shown that paracetamol in the vacuum has a more stable structure in its negative charge state (-1) with -77.71kcal/mol indicating that it has a tendency to behave as an electron acceptor, geometry optimizations with other environments (Water) are performed, in neutral charge state and ionized with one proton and one electron, to determine if the molecule behaves as an acceptor or electron donor, the results are reported in Table 2, where the first thing you notice is that there is a significant decrease in the value of the Heat of Formation when passing from the vacuum to the aqueous environment which shows a great affinity of paracetamol with water; another important aspect is that in vacuum, paracetamol behaves as an electron acceptor, while in water is neither donor nor acceptor, and at least for this structure the bond mechanism by free radical is discarded.

Table 2. Heat of Formation in kcal/mol for paracetamol for vacuum and water in various charging conditions, results obtained with the method PM3.

Charge State	Heat of formation	
	Vacuum	Water
+1	117.73	-1147.02
0	-67.12	-1409.72
-1	-77.71	-1401.16

Graphical representations of the electrostatic potential, also known as maps of potential energy, allow us to illustrate the distributions of molecular charge in two and three dimensions, these allow us to visualize how they molecular charge is distributed and how electrically charged molecules regions behave, knowing this allow us to determine how molecules interact with each other.

In the case of paracetamol, it can be seen that in the neutral charge state the more reactant sites are those in which nitrogen and oxygen atoms located, when the molecule is ionized, that is, in the state -1 the most reactive regions become larger and tend to cover the aromatic ring all around.

However, the state where the molecule has more reactivity is the corresponding to the ionization +1, it means, in the presence of a proton; in this case the entire molecule is reactive, the Heat of Formation values for this state shown that the molecular system at the charge state can exist in the aqueous environment (Figure 4). The best reactivity, observed through the graphical representations of electrostatic potential, shows that the more reactive places correspond to the nitrogen and oxygen atoms, and when ionize an electron, it means in the state -1 the most reactive regions become larger and tend to cover the surroundings of the aromatic ring.

Charge +1 0 -1

Electrostatic Potential

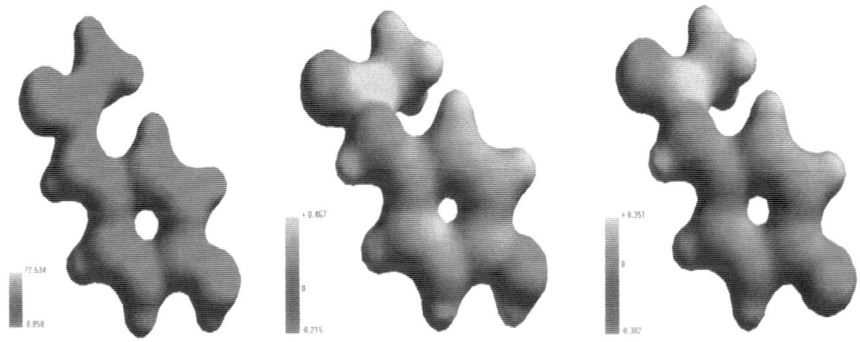

Figure 4. Graphical representation of electrostatic potential, for paracetamol in various states of ionization. Results obtained from the application of ZINDO / CI method.

From previous studies it is known the acceptor behavior of the electron possessing the lignosulfonates [7]. The optimization results of paracetamol in ammonium environment attached from one to three lignosulfonates units in vacuum and water are presented in Table 3.

Table 3. Heat of Formation in kcal/mol for paracetamol in combination with the LS studied in vacuum with ammonium environment and various Charging states. Values obtained using AM1 and PM3 semiempirical methods.

Paracetamol + LS with:	Charge	1 Unit	2 Units	3 Units
	+1	-108.79	-305.67	-502.14
Ammonium in vacuum	0	-287.27	-484.87	-683.88
	-1	-332.97	-532.37	-739.16
	+1	-2431.56	-2751.04	-3066.47
Ammonium in water	0	-2598.47	-2925.17	-3238.89
	-1	-2655.64	-2972.61	-3298.66

The data for in the Table 3 shows the values of the Heat of Formation for new molecular system, this value show the affinity of paracetamol with the environment considered in this work (Ammonium) and lignosulphonates growth, allowing us to visualize the molecules stability. These results, in vacuum and aqueous environments exhibit a high affinity for the paracetamol molecule with aqueous environments.

In these new molecular systems, the LS were incorporated as they grow, and the macromolecules properties can be inferred with those corresponding to small polymeric precursors and fragments.

In the Figure 5 we present the graphical representations of the geometry molecular of the studied systems show that the alignment phenomenon actually occurs, just that it is not as significant as had been observed for natural lignin, the combination with ammonium does not presents the alignment phenomenon, but in the aqueous environment the LS closed around of paracetamol entrapping inside.

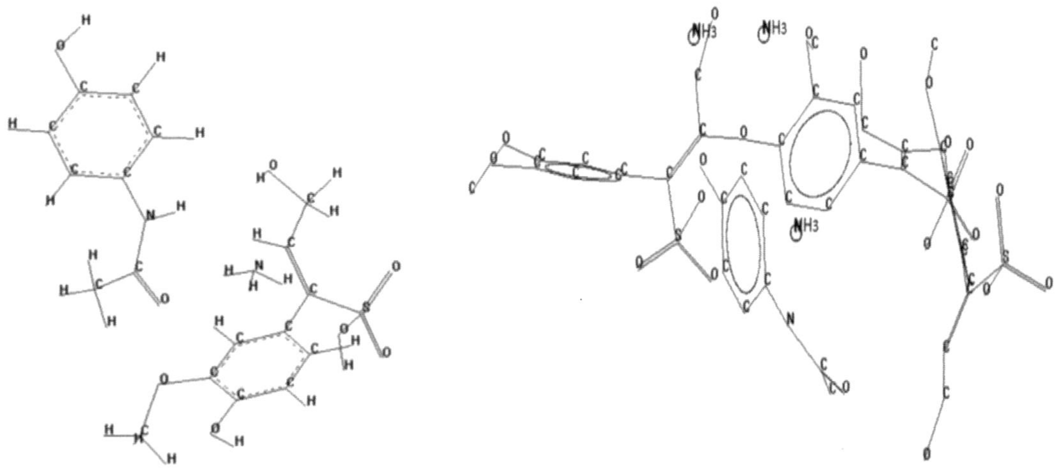

(a) (b)

Figure 5. Graphic representation of the molecular interactions between paracetamol and monomeric LS fragments studied in water. (a) In vacuum, (b) in aqueous ammonium.

4. Conclusions

The results from the calculations carried out with the paracetamol, on the environments considered show that they have a great compatibility with those that contain water, particularly if molecules of an acid are present. The paracetamol also behaves as an electron acceptor showing a great tendency to interact with other molecular systems by the free electron mechanism and also showing a hint of the reasons for it´s great pain relief action.

The electric and optical properties calculated also have shown us that paracetamol has a variety of regions with high electronegativity, and that the size of this regions is related to the environment were the molecule is set up, with acid environments being the most reactive ones, it is also found the LS studied also share the same electric and optical characteristics.

By studying the interaction between the paracetamol and LS molecules (LS with ammonium) we have found that, as expected, there is an interaction with the potential of becoming a "molecular capture" mechanism. We have seen that in the molecular structure of the resulting systems the LS "embraces" the paracetamol with the heat of formation values getting smaller as the geometry of the systems are optimized. Besides the addition of the aqueous medium assists in the process by lowering both, heat of formation and distance between molecules with an aligning of the aromatic parts and the interlacing of the carbon chains.

All this characteristic show us that the LS have a great potential to be used as aids in the removal of paracetamol molecules from the residual waters. The great variety of LS existing in the market, it´s low prices combined with the fact that they are very environmentally friendly gives this alternative a great deal of attractiveness.

Acknowledgments

The authors desire to thank the Vicerrectoria de Investigaciones of the Univesidad del Cauca for the support in the development of this work under the research project ID4064.

References

[1] Bales J R, P J Sadler, J K Nicholson and J A Timbrell 1984 Urinary excretion of acetaminophen and its metabolites as studied by proton NMR spectroscopy *Clin Chem* **30** 1631-1636

[2] James J P Stewart 1989 Optimization of parameters for semiempirical methods I Method *Journal of Computational Chemistry* **10** 209

[3] Eero Sjöström 1993 *Wood Chemistry Second Edition* (San Diego: Academic Press) chapter 4 pp 71-89 doi:10.1016/B978-0-08-092589-9.50008-5

[4] Forrest J A, Clements J A, Prescott L F 1982 Clinical pharmacokinetics of paracetamol *Clin Pharmacokinet* **7(2)** 93-107

[5] Gisela Richardson, Yemin Sun, Maud Langton, Anne-Marie Hermansson 2004 Effects of Ca - and Na-lignosulphonate on starch gelatinization and network formation *Carbohydrate Polymers* **57** 369–377

[6] John Ralph, Brunow Gösta and Boerjan Wout 2007 *Lignins Encyclopedia of life sciences* (UK: John Wiley & Sons Ltd) pp 1-10 doi: 10.1002/9780470015902.a0020104

[7] P J Salazar Valencia, L E Bolívar Marinez, S T Pérez Merchancano 2015 Molecular modeling of ammonium, calcium sulfur and sodium lignosulphonates in acid and basic aqueous environments *Brazilian Journal of Physics* **45(6)** 567-574

[8] J Paul Simon, Karl-Erik L Eriksson 1996 The significance of intra-molecular hydrogen bonding in the β-O-4 linkage of lignin. *Journal of Molecular Structure* **384** 1-7

[9] P J Salazar-Valencia, S T Pérez-Merchancano and L E Bolívar-Marinéz 2006 Optical Properties in Biopolymers Lignin Fragments *Brazilian Journal of Physics* **36(3B)** 840-843

[10] Hypercube 2003 *HyperChem Professional 7.51* (Florida: Hypercube, Inc.) http://www.hyper.com/

Journal of Physics: Conference Series **743** (2016) 012012 doi:10.1088/1742-6596/743/1/012012

Frozen magnetic response in mesoscopics superconductors

F Durán Flórez[1], E D V-Niño[2,4] and J Barba-Ortega[3,4]
[1] Universidad Pontificia Bolivariana, Bucaramanga, Colombia.
[2] Universidad Politécnica de Madrid, Madrid, España.
[3] Universidad Nacional de Colombia, Bogotá, Colombia.
[4] Foundation of Researchers in Science and Technology of Materials, Colombia.

E-mail: jjbarbao@unal.edu.co

Abstract. For a bulk type II superconducting sample at low temperature, the magnetic field can penetrate in the form of a single quantum fluxoid, for bulk samples this fluxoides are arranged in a hexagonal lattice, this so-called Shubnikov-Abrikosov state or vortex state and takes place between the first and the second critical thermodynamics magnetic fields. Under the first magnetic critical field is present the Meissner-Oschenfeld state. For mesoscopic samples, the magnetic response can present very interesting properties due the proximity effect of in-homogeneous boundary conditions and the presence of dots, anti-dots and/or impurities. The superconducting state in a mesoscopic sample with dot/anti-dot/dot/pillar/trench is calculated within the nonlinear Ginzburg-Landau equations. We predict that the critical magnetic fields and magnetization depends on strongly of the nature, geometry, size of the defects and the boundary conditions used.

1. Introduction

The superconducting-magnetic properties of samples with magnetic-structured-topological pinning or anti-pinning cores have attracted over the past years [1, 2]. The present possibility of controlling the magnetic and physics properties of the superconducting condensate (vortex state, critical fields and currents) has made from superconductors of size of the coherence or penetration length the principal experimental and theoretical systems for studies [3,5]. Is known that when the magnetic-structural-topological defects are placed into the sample, the vortex matter depends on their configuration, density and the pinning array [2]. In previous works one of the authors of this paper studied the superconducting properties of a Niobium prism with its lateral surfaces in contact with deferent kinds of metallic and/or superconducting materials, the authors found that the second thermodynamic field increase when a superconducting-superconducting interface is considered and a slow entry of the magnetic field is observed in the metallic regions of the boundary [6]. In addition, they studied the influence of several thermal gradients on the vortex configuration and the thermodynamics properties of a low critical temperature nanoscopic superconducting long square prism [7]. The field for the first vortex penetration presents a slow dependence on the exponent of the gradient and the size of the sample [7]. The magnetic susceptibility of a superconductor immersed in a magnetic field is calculated numerically using the Ginzburg-Landau model; the authors found that the magnetic susceptibility shown a quasi-periodic modulation at the vortex transition fields [8]. Baelus found that increasing

Content from this work may be used under the terms of the Creative Commons Attribution 3.0 licence. Any further distribution of this work must maintain attribution to the author(s) and the title of the work, journal citation and DOI.
Published under licence by IOP Publishing Ltd

Journal of Physics: Conference Series **743** (2016) 012012 doi:10.1088/1742-6596/743/1/012012

field, the number of vortices simultaneously entering the sample depends on the presence and the positions of the surface defects [9]. Now, in this work we solved the time dependent Ginzburg-Landau equations for calculating the magnetic response in several mesoscopics superconducting samples with anti-dot/dot/pillar/trench. We found a strong/weak dependence of the critical fields and magnetization on the size, boundary conditions, geometry and natures of these defects.

2. Theoretical Method

The time dependent Ginzburg-Landau (TDGL) equations for isotropic superconductors, which govern the superconducting order parameter and the vector potential in the zero electric potential gauge, are given by [10-14]:

$$\frac{\partial \Psi}{\partial t} - D^2\Psi - \Psi^3 + \Psi = 0 \tag{1}$$

$$\vec{J}_T = -\frac{\partial \vec{A}}{\partial t} = \vec{J}_S + \vec{J}_n \tag{2}$$

Where $D = i\nabla + \vec{A}$, $\mathbf{J_T}$ is the total current, $\vec{J}_n = k^2\nabla \times \nabla \times \vec{A}$ is the normal current, and $-\mathrm{Re}(\overline{\Psi} D\Psi) = \vec{J}_s$ is the super-current. The time dependent Ginzburg Landau equations were rescaled as follows: Ψ in units of $\Psi_\infty(0)$, lengths in units of the coherence length $\xi(0)$, the external applied magnetic field H_e in units of the second critical thermodynamics field in the bulk $H_{c2}(0)$, potential vector A in units of $H_{c2}(0)\xi(0)$, temperatures in units of the critical temperature T_C. The TDGL equations are complemented with the general boundary conditions for the order parameter: $D\Psi = \Psi/b$, b is deGennes extrapolation length and simulate the interface with another material. The magnetization $-4\pi\vec{M} = \vec{B} - \vec{H}_e$ (here **B** is the induction, the spatial average of the local magnetic field) is [15]:

$$-4\pi M_n = \frac{1}{N_x N_y}\sum_{i=1}^{N_x}\sum_{j=1}^{N_y} h^n_{z,i,j} - H_e \tag{3}$$

3. Results and discussion

In this work, we report the dependence of several kinds of defects on the magnetization and thermodynamics critical fields in a superconducting nanoscopic flat disk and parallelepiped submerged in an applied magnetic field. In thin disk limit, we considered a superconducting disk of Nb of radius R = 1 μm, and thickness d ≈ 45 nm at T = 0. The TDGL equations are solved numerically in two different scenarios. One is that in which the system has invariance along the z-direction in which the external field is applied; this is the case of a circular cross section. In this scenario, we have a bi-dimensional system and demagnetization effects are not present, that is, the local magnetic field outside the sample is equal to the external applied magnetic field [16]. The second scenario is the thin film limit, for which the first TDGL equation (for the order parameter) is solved only at the plane perpendicular to the direction of the applied magnetic field, and the second equation (for the potential vector) is solved by substituting the super-current density with a superconducting sheet current density. The Studied cases are:

- Case 1: A triangular and circular central dot (anti-dot) in a thin disk.

- Case 2: A pentagonal and hexagonal central defect at higher (lower) critical temperature in a disk.
- Case 3: A pentagonal and hexagonal trench (barrier) in a thin disk with a small central perforation.
- Case 4: A mesoscopic parallelepiped (considering a bi and three dimensional cases) with a square pillar on its top surface.

The magnetic induction in the sample for cases 1, 2 and 3, is considered equal to an external applied magnetic field, so the second TDGL equation not is solved and $A = A_0$. For bi (and not thin) and three-dimensional samples a complete set of the TDGL equations must be solved (case 4).

3.1. Case 1

For a very thin disk of variable thickness, the TDGL equations can be rewritten as:

$$\frac{\partial \Psi}{\partial t} = \frac{1}{\mu}\left(i\nabla + \vec{A}_0\right) \bullet \mu\left(i\nabla + \vec{A}_0\right)\Psi + |\Psi|^3 - \Psi \tag{4}$$

$\mu(r,\theta)$ is a function which describes the thickness of the sample. We considered $\mu(r,\theta) = 1$ everywhere, except at dot/anti-dot position in the disk which is simulated by using $\mu(r,\theta) = 0.85$ for the dot and $\mu(r,\theta) = 1.25$ for the anti-dot. Initially, we analysed the first entrance of vortices into the sample, for disk with circular and triangular defects (dot and anti-dot) and thin disk without defect, we found that all samples exhibit the first jump of the magnetization at equal magnetic field, for the second vortex entry, we found differences in the location of peaks for different values of applied magnetic field, the vorticity decreases when the anti-dot is used, this fact is due to the anti-dot generate greater opposition to the vortices entrance, while the opposite occur with the dot [17]. The vortices always (never) are located into the dot (anti-dot) thus, the pinning force is present in the dot and the force of vortex–vortex interaction overcomes the repulsive force exerted by the anti-dot. The superconductor/normal transition field or second critical field is the same for all studied samples.

3.2. Case 2

Thin disk with regions on its surface at different critical temperature described by the function $\mu(r,\theta)$. For this case, the first TDGL can be rewritten as:

$$\frac{\partial \Psi}{\partial t} = \left(i\nabla + \vec{A}_0\right)^2\Psi + \Psi(\mu - \Psi^2) \tag{5}$$

We considered $\mu(r,\theta) = 1$ everywhere, except for $\mu(r,\theta) = 0.85$ for the superconductor at lower critical temperature and $\mu(r,\theta) = 1.25$ for the superconductor at higher critical temperature. In our analysis, we found that the first transition field is independent of the defect geometry and the kind of the defect used. We found differences in the magnetization curve when different kinds of defects are studied at low magnetic applied fields, but the second critical field does not depend on the geometry and nature of the defect. In another hand, in the down branch of the magnetic field at zero magnetic fields, the magnetization is zero for the hexagon case, while for a pentagonal defect into the pentagon defect is different to zero (paramagnetism).

3.3. Case 3

In this case, the used TDGL equation is the same equation used for the case 1. Here, we analysed the influence of the boundary conditions on the magnetization and critical magnetics fields. The general boundary conditions for a superconductor are $D\Psi_s = \Psi_s/b$ (b is the extrapolation deGennes length). We will assume either a superconductor/vacuum external interface ($b \to \infty$), and the surface of the sample in a completely normal state (b = 0). We found that the second critical field is independent of the geometry and height of the defects in the disk [18]. Moreover, the first vortex penetration field depends on the pinning or anti-pinning force of the defects; this field is lower for a barrier in both geometries. This means that the barrier repeals the first penetration than the trench, which was expected, but surprisingly, it does not depend on the geometry used [18]. For both pentagonal and hexagonal cases, the second critical field occurs at the same value. The first vortex penetration field and second critical fields dependent on the boundary condition. These fields are equals when we use $b \rightarrow \infty$ for all cases, whereas if we take b = 0, we have the first field is lower for the trench case that for barrier case, also the second field is the same for both geometries. If we compare these results with previous works [19-20], we found that the first and second critical fields for a superconducting disk with a square, circular or triangular trench are independent of the shape of the defects and depends strongly on the boundary conditions.

3.4. Case 4

A parallelepiped with a square pillar on its top surface: The geometry of the problem includes the demagnetization effects, so the magnetic field in the surface is higher that the applied magnetic field. The superconducting domain covers the parallelepiped of thickness c and lateral sizes a and b. The dimension of no superconducting and superconducting region is A ×B ×C. The pillar dimensions are W×W×a, the first and second GL can be rewritten as:

$$\frac{\partial \Psi}{\partial t} = \left(i\nabla + \vec{A}\right)^2 + |\Psi|^3 - \Psi + i\left(-i\nabla - \vec{A}\right)\bullet\frac{\nabla\mu}{\mu} \tag{6}$$

$$\vec{J}_T = -\frac{\partial \vec{A}}{\partial t} = \vec{J}_S + \vec{J}_n \tag{7}$$

$J_s = 0$ out of the superconducting area and just the normal current is taken in count [16]. Again, we considered $\mu(r,\theta) = 1$ everywhere, except $\mu(r,\theta) = 1.25$ for the pillar. We analyzed the diamagnetic response of the superconductor both from the three and bi dimensional cases. We found that the first penetration field and the field corresponding to the maximum of magnetization depend strongly on the size and dimensionality of the sample, being bigger for bi-dimensional samples. One difficulty in making a comparison between the bi and three models on equal footing comes from the fact that most of the configurations with equal vorticity belong to distinct values of the external applied magnetic field. This is so, because in the bi-dimensional model, the demagnetization effect not is considered and requires a larger applied field to enhance the shielding currents and consequently produces the same picture observed in the three dimensional model at a larger value of the external magnetic field [16]. We have found that the presence of the pillar perturbation diminishes the difference of the first penetration field from approximately the half between them [21].

4. Conclusion

In this contribution, we analysed the magnetic response of a mesoscopic superconducting sample (disk and parallelepiped) with several kinds of internal structural defects (dots/anti-dot/trench/pillar) and different boundary conditions in presence of an external homogenous magnetic field, by solving numerically the 2D and 3D non-linear Ginzburg–Landau differential equations. Our results show that magnetic properties of all studied samples depend on the dimensions, nature, the geometry of the imperfections, and boundary conditions used in each system. The small size of the sample (mesoscopic regime) determines strongly the influence of the defects on the critical fields and magnetization. Also, the pinning (anti-pinning) force of these defects determines the surface energy barrier and subsequently the magnetic field for the first vortex penetration.

Acknowledgments
The authors thank Edson Sardella by useful discussions.

References

[1] J Van de Vondel, C C de Souza, B Y Zhu, M Morelle and V V Moshchalkov 2005 Vortex-Rectification Effects in Films with Periodic Asymmetric Pinning *Phys. Rev. Lett.* **94** 057003

[2] K Yu, T W Heitmann, C Song, M P DeFeo, B L T Plourde, M B S Hesselberth and P H Kes 2007 Asymmetric weak-pinning superconducting channels: Vortex ratchets *Phys. Rev. B* **76** 220507

[3] K Yu, M B S Hesselberth, P H Kes and B L T Plourde 2010 Vortex dynamics in superconducting channels with periodic constrictions *Phys. Rev. B* **81** 184503

[4] M B Hastings, C J O Reichhardt and C Reichhardt 2003 Ratchet Cellular Automata *Phys. Rev. Lett.* **90** 247004

[5] T Puig, E Rosseel, M Baert, M J V Bael, V V Moshchalkov and Y Bruynseraede 1197 Stable vortex configurations in superconducting 2×2 antidot clusters *Appl. Phys. Lett.* **70** 3155

[6] J Barba-Ortega, C A Sjogreen and Miryam R Joya 2015 Weak lineal variation of the Neumann boundary conditions in a superconducting plate *Universidad, Ciencia y Tecnologia* **19(77)** 155

[7] J Barba-Ortega, H Carrillo and C A Sjogreen 2016 Unidimensional thermal gradients Tn in a nanoscopic superconductor *Physica C* **525** 61

[8] C A Aguirre, J D Gonzalez and J Barba-Ortega 2016 Effects of Size, deGennes and Ginzburg–Landau Parameters on the Magnetic Susceptibility of an Isotropic Superconductor *J. Low Temp. Physics* **182(1)** 51

[9] B J Baelus, K Kadowaky and F M Peeters 2005 Influence of surface defects on vortex penetration and expulsion in mesoscopic superconductors *Phys. Rev. B* **71** 024514

[10] P J Pereira, V V Moshchalkov and L F Chibotaru 2014 Efficient solution of 3D Ginzburg-Landau problem for mesoscopic superconductors *J. Phys. Conf. Ser.* **490** 012220

[11] D Gropp, H G Kaper, G K Leaf, D M Levine, M Palumbo, V M Vinokur 1996 Numerical simulation of vortex dynamics in type-II superconductors *J. Comput. Phys.* **123** 254

[12] G C Buscaglia, C Bolech and A López 2000 *Connectivity and Superconductivity* ed J Berger and J Rubinstein (Heidelberg: Springer) p 200

[13] P G de Gennes 1994 *Superconductivity of Metals and Alloys* (New York: Addison-Wesley) p 171

[14] M V Milosevic and R Geurts 2010 The Ginzburg-Landau Theory in application *Physica C* **470** 791

[15] Sindy Jessenia Higuera Agudelo, Heli Barba Molina & José José Barba Ortega 2014 Effect of a superconducting defect on the Cooper pairs of a mesoscopic sample *DYNA* **81(186)** 108-112

[16] J Barba-Ortega, E Sardella, J Albino Aguiar 2015 Superconducting properties of a parallelepiped mesoscopic superconductor: A comparative study between the 2D and 3D Ginzburg–Landau models *Phys. Lett. A* **379** 732

[17] J D Gonzalez, Miryam R Joya and J Barba-Ortega Influence of short-range artificial defects in a macroscopic flat disk on the Abrikosov state *International Journal of Modern Physics B* **28(1450227)** 1-13

[18] J Barba-Ortega, J D González, Edson Sardella 2014 Superconducting State of a Disk with a Pentagonal/Hexagonal Trench/Barrier Journal of Low Temperature Physics **174(1)** 96–103

[19] J Barba-Ortega, E Sardella, J Albino Aguiar 2013 Superconducting state of a perforated mesoscopic disk with a sqare or triangular trench *Mod. Phys. Lett. B* **27(15)** 1350115

[20] J Barba-Ortega, E Sardella, J Albino Aguiar 2013 Nucleation of superconductivity in a thin disk with a ring-like defect *Mod. Phys. Lett. B* **27(4)** 1350025

[21] F Rogeri, R Zadorosny, P N Lisboa-Filho, E Sardella and W A Ortiz 2013 Magnetic field profile of a mesoscopic SQUID-shaped superconducting film *Sup. Sci. Tech.* **26** 7

Journal of Physics: Conference Series **743** (2016) 012013 doi:10.1088/1742-6596/743/1/012013

Quantitative phase analysis from powder diffraction using de Rietveld method in hydrogen storage alloys based on TiCr

A Martinez[1,2], D Bellon[1] and L Reina[1]
[1] Universidad de Santander, Bucaramanga, Colombia.
[2] Foundation of Researchers in Science and Technology of Materials, Colombia.

E-mail: alejandrom@udes.edu.co

Abstract. Hydrogen storage is one of the important steps in the implementation of the hydrogen economy; metal hydrides are a promising way to achieve this goal. We present in this work the use of Rietveld analysis to characterize structurally TiCr-based alloys that are able to store hydrogen. $TiCr_{1.1}V_{0.9}$, $TiCr_{1.1}V_{0.45}Nb_{0.45}$, $TiCr_{1.1}V_{0.2}Nb_{0.8}$, $TiCr_{1.1}Nb_{0.9}$ alloys were synthesized in an arc furnace under argon atmosphere. The analysis of phases was developed by X-Ray Diffraction (XRD) for further refinement of both the two lattice parameters and the percentage of the phases. Our results confirmed that a structure bcc, mostly combined with a small percentage of Laves phases, leads to obtain important properties in this area. *Rietveld* analysis was performed by the *Fullprof* program and this program allows us to obtain the different structural parameters.

1. Introduction

Hydrogen storage is one of the keys to implement a new global energy vision based on this gas, the methods currently used are pressurized tanks, cryogenic tanks and metal hydrides [1]. The last one has advantages for its large storage capacity and safety, and for these are used alloys with transition metals such as Mg which stores 7% by weight of H_2 and metals such as Ti, Cr, V, Zr, Nb [2]. Hence the great interest in studying the crystal structure and physic-chemical properties to improve performance in the hydrogen absorption capacity. The X-ray diffraction is one of the most important techniques for characterization of crystalline materials. This has traditionally been used for qualitative analysis, quantitative phase and for the determination of the crystal structure. In recent years their applications have been extended to the determination of imperfections in the structure and residual stress measurements. The *Rietveld* method is a refinement technique that has the ability to determine with accurately structural parameters a crystalline sample, from the construction of a theoretical model that fits at the experimental diffraction pattern by the least squares method. In the theoretical model are included structural aspects such as crystal structure, space group, position of atoms in the unitary cell, etc. Also, in the theoretical model are included microstructural factors that consider the concentration of the phases present, crystal size and microstrains. The instrumental factor is also included, which contemplates the effect of optical diffraction equipment X-ray on the measurement and whose effect is the broadening of the diffraction peaks [3].

In this paper is studied the effect on the lattice parameters and on the quantification of the phases by replacing V by Nb in based alloys on TiCrV used for hydrogen storage, by the *Rietveld* method by using the *fullprof* program.

Content from this work may be used under the terms of the Creative Commons Attribution 3.0 licence. Any further distribution of this work must maintain attribution to the author(s) and the title of the work, journal citation and DOI.
Published under licence by IOP Publishing Ltd

2. Experiment methodology

Button shaped ingots of TiCr1,1$V_{0,9}$, $TiCr_{1,1}V_{0,45}Nb_{0,45}$, $TiCr_{1,1}V_{0,2}Nb_{0,8}$, $TiCr_{1,1}Nb_{0,9}$ alloys with weights of 20g were prepared by arc melting in argon atmosphere, using high-purity materials (>99.95). The samples were melted 3 times in order to achieve a homogenized microstructure.

For structural study was necessary to use the sample metal powder which passed through a mesh of 270 microns, following which it was subjected to XRD using a Shimadzu 5000 with Cu-Kα radiation, the lattice parameters and phase present in the samples were calculated by the *Rietveld* method using the software *Fullprof.*

3. Analysis and discussion of results

Figure 1 shows X-ray diffraction patterns for the alloys. Figure 1(a) shows a typical bcc structure type (Im3m space group) for the $TiCr_{1.1}V_{0.9}$ alloy [4]. As is known in the TiCr binary phase diagram, the stoichiometric ratio $TiCr_{1.1}$ crystallizes in Laves phase [5].

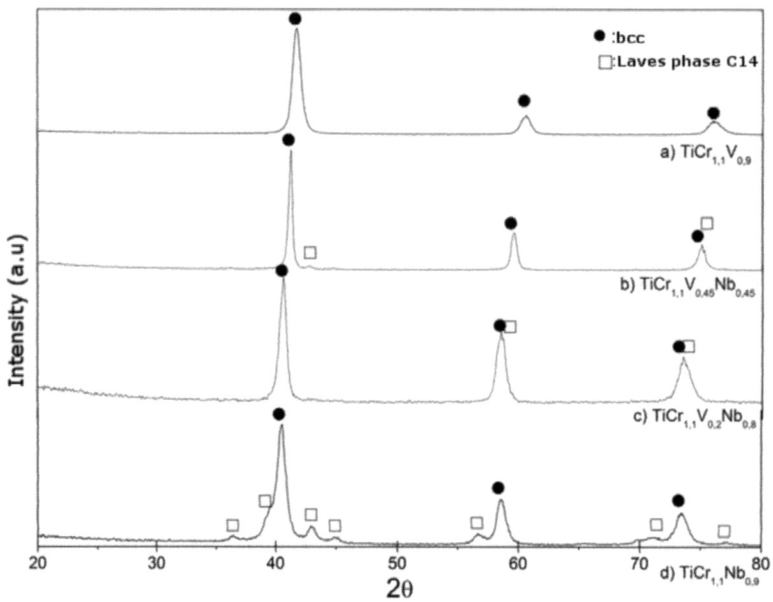

Figure 1. X-ray diffraction patterns of the different alloys, (a) $TiCr_{1.1}V_{0.9}$, (b) $TiCr_{1.1}V_{0.45}Nb_{0.45}$, (c) $TiCr_{1.1}V_{0.2}Nb_{0.8}$, (d) $TiCr_{1.1}Nb_{0.9}$.

When an element like Vanadium (V) is mixed, the inhibition of the formation of Laves phase in the melting process is presented [6], producing a bcc structure, as is observed in figure 1.a. It is also noted that the V inhibits the increase of Laves phase when Niobium (Nb) is added. However, when the Nb content is increased and the V content decreased, the amount of Laves phase C14 increases, coexisting with the bcc structure as it is shown in Figures 1(b) and 1(c). When V is completely replaced by the Nb, the Laves phases raised significantly.

The Laves phases are presented in compounds or alloys where the ideal ratio of atomic sizes between neighboring atoms A and B is $\sqrt{(3/2)}=1.225$; however, experimentally they are in a range of solubility between 1.06 and 1.67. This feature is presented by the geometric factor of atoms due to the property of contract and expand so that approximates the ideal ratio [7]. Therefore, the geometric rule is important for predicting behaviors of Laves phases.

In this work is expected that the $TiCr_{1.1}Nb_{0.9}$ alloy crystallizes a Laves phase, since each pair of elements fit the ideal reason ≈1,225 (R_{Nb}/R_{Cr}=1.15 and R_{Ti}/R_{Cr}=1,14). Thus, the ternary alloy TiCrNb

Journal of Physics: Conference Series **743** (2016) 012013 doi:10.1088/1742-6596/743/1/012013

meets the geometric principles for the formation of Laves phases coexisting with a bcc structure, as observed in the diffractogram Figure 1(d). This is expected in the transition metals due to the relative similitude of the atomic radii of the metal elements (R_{Nb}=1.468Å, R_{Ti}=1.462Å, R_{Cr}=1.282Å).

It is observed an increase in lattice parameters for a bcc structure, with high intensive to lower angles. The lattice parameters and the percentage of the phases present are calculated by the *Rietveld* method using the *fullprof* program, and are the results are presented in Table 1. The graphics of *Rietveld* analysis are shown in Figure 2.

Table 1. Rietveld analysis results for the calculate the lattice parameters and the percentage of the phases present in the different alloys.

Sample	S	R_{pw}	Phase	% Phase	Lattice Parameters (Å)
$TiCr_{1.1}V_{0.9}$	1.40	12.36	bcc	100.00	a=3.06
$TiCr_{1.1}V_{0.45}Nb_{0.45}$	1.78	28.08	bcc	99.50	a=3.10
			C14	0.50	a=4.94, b=8.15
$TiCr_{1.1}V_{0.2}Nb_{0.8}$	1.75	22.40	bcc	97.50	a=3.15
			C14	2.50	a=5.18, b=9.56
$TiCr_{1.1}Nb_{0.9}$	5.87	27.20	bcc	85.00	a=3.15
			C14	15.00	a=4.90, b=8.14

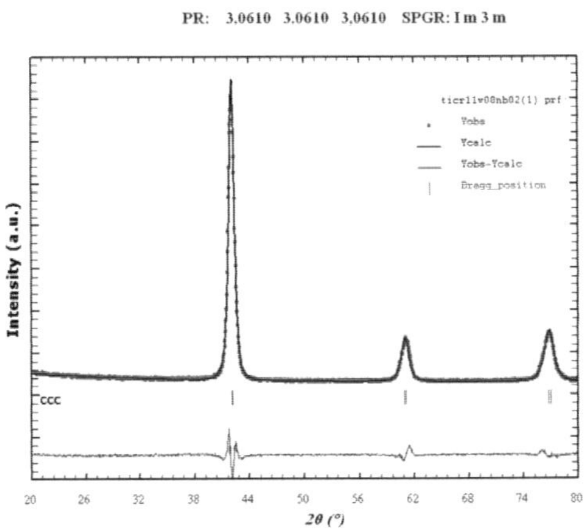

(a)

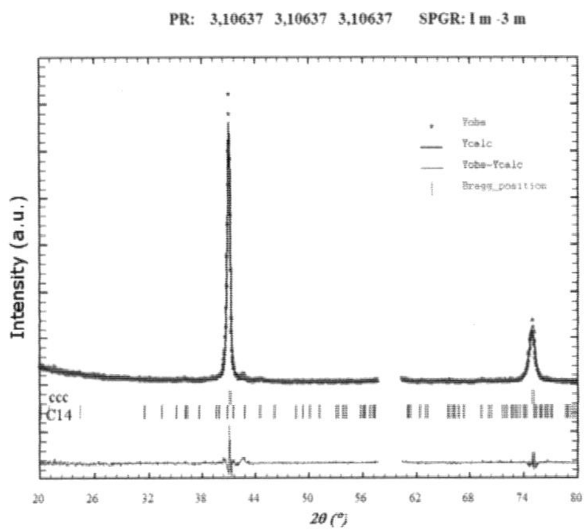

(b)

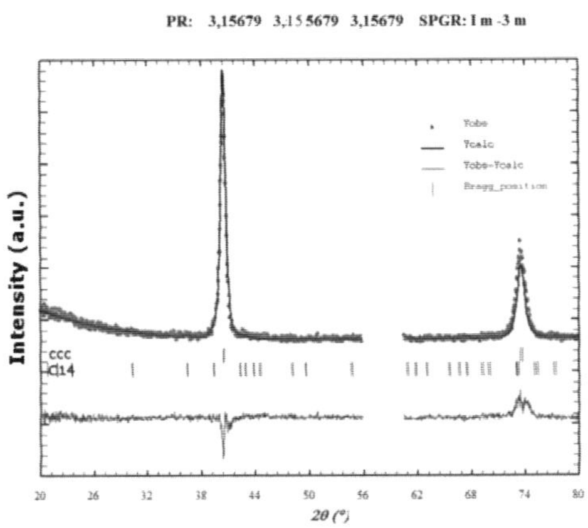

(c)

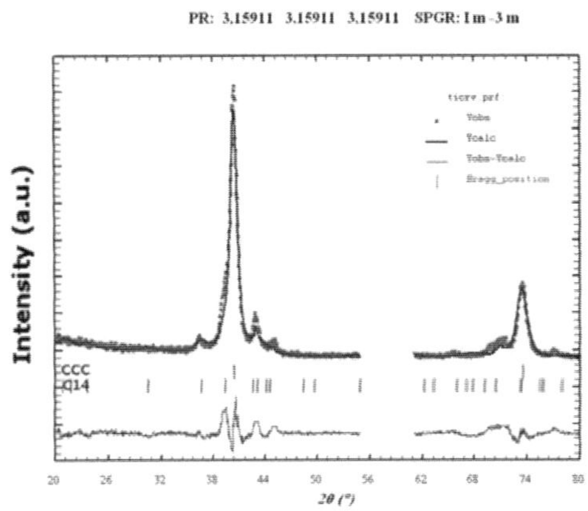

(d)

Figure 2. Rietveld Analysis of the samples (a) $TiCr_{1,1}V_{0,9}$, (b) $TiCr_{1,1}V_{0.45}Nb_{0.45}$, (c) $TiCr_{1,1}V_{0.2}Nb_{0.8}$, (d) $TiCr_{1,1}Nb_{0,9}$.

The *Rietveld* method is based on the relationship:

$$S_y = \sum W_i\left(y_{i(obs)} - y_{i(cal)}\right)^2 \tag{1}$$

In this function the $y_{i(obs)}$ and $y_{i(cal)}$ are the experimental and calculated intensity at point i of the diffraction pattern respectively, W_i is the respective weight given to these intensities while the summation is over all points of the pattern diffraction. One way of following the progress of the refinements is naturally to watch the decrease of this least-squares residual. In most cases the residual value is normalized and expressed as:

$$R_{pw} = \left(w_i \sum\left(y_{i(obs)} - y_{i(cal)}\right)^2 / \sum w_i (y_{i(obs)})^2\right)^{1/2} \tag{2}$$

The values shown in Table 1 have a good approximation of the quantification of the phase present in each alloy and the calculation of the lattice parameters. The discontinuity in Figure 2 of the *Rietveld* analysis does not affect the measure because of the convergence of the algorithm.

Since the number of atoms and atomic radii of Ti (1.462Å) and Nb (1.468Å) are larger than atoms and atomic radii of Cr (1.282Å) and V (1.34Å), the lattice parameter increases for the different samples. This leads to an increase in the lattice parameter, as far that vanadium is substituted for niobium.

The fact that the unit cell of the bcc structure is increased is supported by the empirical *law of Vegard*, which describes the increase or decrease linear lattice parameter as a function of the concentration of the constituent elements of the alloy. Yoo *et al.* [8] reported this phenomenon, where they observed a reduction of network parameters when they were added by small percentages of Cr in the Fe alloy $Ti_{0.32}Cr_{0.43}V_{0.25}$.

4. Conclusions

Substituting V for Nb in the alloy $TiCr_{1.1}V_{0.9}$ was studied. In this work we verified that the compounds $TiCr_{1.1}V_{0.45}Nb_{0.45}$, $TiCr_{1.1}V_{0.2}Nb_{0.8}$ and $TiCr_{1.1}Nb_{0.9}$ give rise to a double-phase structure made up of primarily a bcc structure and a smaller Laves C14 cubic-type structure.

Substituting the V by Nb the lattice parameters increases this is a good indicator for the type of alloys from hydrogen storage.

References

[1] Louis Schlapbach and Andreas Zuttel 2001 *Nature* **414** 390
[2] Andreas Zuttel 2003 *Materialstoday* **6** 33
[3] Talhita C. Santini 2015 *International Journal of Mineral Processing* **139** 10
[4] A Okada M, Kuriiwa T, Tamura T, Takamura H, Kamegawa A 2002 *Journal of Alloys and Compounds* **330(332)** 511
[5] Kabotumori T, Takeda H, Wakisaka Y, Ohnishi K 1995 *Journal of Alloys and Compounds* **231** 528
[6] D S dos Santos, M Bououdina, D Fruchart 2002 *Journal of Alloys and Compounds* **340** 101
[7] F Stein, M Palm, G Sauthoff 2004 *Intermetallics* **12** 713
[8] Jeong-Hyun Yoo, Gunchoo Shim, Jae-Sik Yoon, Sung-Wook Cho 2009 *International Journal of Hydrogen Energy* **34** 1463

Simulation of stress concentration factors in combined discontinuities on flat plates

A Santos[1], R Guzman[1], Z Ramirez[2] and C Cardenas[2]
[1] Universidad Pontificia Bolivariana, Bucaramanga, Colombia.
[2] Unidades Tecnológicas de Santander, Bucaramanga, Colombia.

E-mail: alfonso.santos@upb.edu.co

Abstract. The objective of this paper was to determine the stress concentration factors for static load conditions (Kc) through the use of a finite elements software. The concentrators originate an increase in the calculated stress to design mechanical elements so that quantify this increase allow to obtain viable designs capable to endure the loads to which they are subjected. Among the most relevant results is validation, through statistical analysis, of use of ANSYS software for the determination of this parameter in flat plates with a central hole subjected to axial load conditions; the graphs of Kc for flat plates with combined discontinuities, central hole – groove and central hole – fillet, under axial load conditions were also obtained; the Kc value for flat plates with central hole – groove has a polynomial relation of degree 5 with the r/h parameter and decreases in so far as that this magnitude increases; additionally is also demonstrated that the concentration factor depends on the H/h relation and there is no defined trend between these two parameters. The Kc value for flat plates with central hole – fillet has a polynomial relation of degree 5 with the r/h parameter and also depends on the relation H/h without showing a defined trend between these two parameters.

1. Introduction

The distribution of stresses through a mechanical element when is subjected to axial loads, torsional or bending moments, is drastically affected by the presence of geometric discontinuities included in the design of parts or present, due to microscopic irregularities proper of manufacturing processes or material defects. The presence of these irregularities results in an increase of stress in the vicinity of the discontinuity such, that in order to obtain pieces capable of supporting the loads applied, these concentration factors must be included in the design process.

The first mathematical study of the stress concentration was published shortly after 1900; in order to work with different simple cases, it has been developed experimental methods to quantify the local stresses and from these data to determine the Kc value as the relation between this magnitude and the nominal stress calculated from simple formulas. In recent years there have begun using computer simulations based on finite elements, which allow to make mathematical models of real systems calculation, in an easiest and most economical way to modify than a prototype [1-5].

The information available in the literature is quite limited and is restricted to simple geometric configurations of stress concentrators, such as holes, fillets or grooves; recently, it has been reported more complex configurations where are involved two or more discontinuities located on different cross sections of flat plates [3, 6]. To validate the use of a finite elements software for obtaining this parameter in more complex conditions as combined concentrator, i.e. several irregularities present simultaneously on the same cross sections, represent a significant contribution to the design process of

Content from this work may be used under the terms of the Creative Commons Attribution 3.0 licence. Any further distribution of this work must maintain attribution to the author(s) and the title of the work, journal citation and DOI.
Published under licence by IOP Publishing Ltd

Journal of Physics: Conference Series **743** (2016) 012014 doi:10.1088/1742-6596/743/1/012014

mechanical elements, as it will allow to predict, in a better way, the maximum stresses present in the components of a machine due to the action of different loads.

2. Conceptual framework

2.1. Stress concentrators

The stress concentrators are geometrical irregularities that cause an increase in the average stress that should occur in regions near these discontinuities; the relationship between the maximum stress that occurs and the average stress to be presented is defined as stress concentration factor, Kc; which is determined by experimental or analytical methods and presented as graphs for easy interpretation.

A typical example of a stress concentrator is a rectangular bar with a fillet, subjected to tensile load, as shown in Figure 1(a). If the bar is cut in the cross section of the fillet, these stresses will be shown in Figure 1(b); the stress distribution along the cut Surface is not uniform and reaches a maximum value in the vicinity of the fillet, where stresses increase considerably. The maximum value of the stress by those points is found by multiplying the average stress for a Kc stress concentration factor [7, 8].

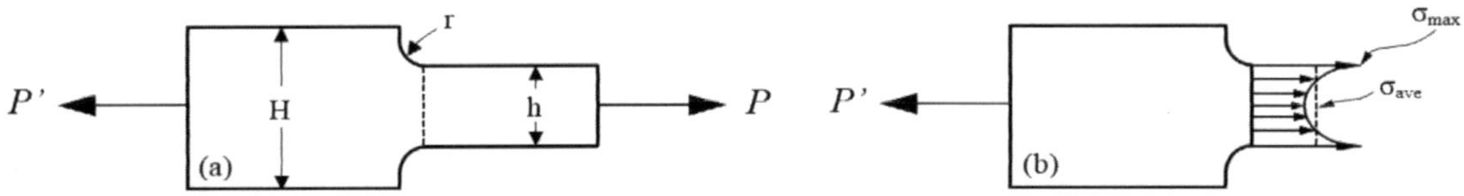

Figure 1. Flat plate with fillet (a) tension load applied and (b) stress distribution in the cross section of the fillet.

2.2. Stress concentration factors, K_c

The stress concentration factor for static load can be determined as the relation of the actual maximum stress in the discontinuity and the average stress, and is obtained through Equation (1) [9].

$$K_c = \frac{Actual\ Maximum\ Stress}{Average\ Stress} \tag{1}$$

The average stresses are determined by the elementary equations and are defined by the type of load that is acting on the element. In the case of an axial load that causes tension or compression, this value is calculated through Equation (2) [10].

$$\sigma_{average} = \frac{Axial\ Force}{Area\ of\ Cross\ Section} \tag{2}$$

The stress concentration factors are function of the discontinuity type, the geometry of the discontinuity and the type of load is experienced.

2.3. Finite elements

The general idea of the finite elements method is the division of a continuum, in a set of small elements interconnected by a series of points called nodes. The equations governing the behaviour of continuum also govern the element. This way is gotten through a continuous system which is governed by a differential equation or a system of differential equations to a system with a finite number of degrees of freedom whose behaviour is modelled by a linear equations system or not [11, 12].

Onto nodes are materialized the fundamental unknowns of the problem; in the case of structural elements, these unknowns are the nodal displacements, as from these can be calculated the remaining unknowns of interest: stresses, strains, etc. The finite elements method as a mathematical formulation is relatively new; although its basic structure has been known for quite some time, in recent years it has undergone a great development due to advances in computer technology. These advances have

Journal of Physics: Conference Series **743** (2016) 012014 doi:10.1088/1742-6596/743/1/012014

been precisely those, which have made available to the user may programs that allow calculations with finite elements.

3. Methodology

The model used for simulation was a flat plate of AISI 1020 steel with two concentrators located on the same cross section subjected to axial load. In order to fulfill with the objective of the project, first is made a validation of the software to use, reason for which the behavior of a sheet with a central hole and subjected to axial load is simulated and the values obtained from the concentration factor are compared with the theoretical data found in the literature for the same condition of stress concentration. Later simulations for conditions of combined discontinuities were made. The methodology for the development of the Project consisted of three stages:

3.1. Simulation of sheets with central hole

The steps carried out to perform the simulations were the following:

- Definition of the model used for the simulations; in this case a static model was defined.
- Entering the mechanical properties of the material to the database software and define it as the default for simulations; the data provided were: modulus of elasticity (E=205GPa), Yield strength (S_y=250MPa) and Ultimate strength (S_{ut}=460MPa).
- Making the graphical representation of the element to simulate; a rectangular sheet of 0.00635m thick, 0.05080m wide and 0.40000m length was drawn; Additionally grooves or fillets were performed in the middle section of the length of the sheet with the dimensions of the radius of the central hole; the relations used between the largest and smallest width of the sheets were 1.1, 1.5 and 2.0.
- Performing meshing process, in Figure 2(a) and 2(b) is shown the meshing result of the element for combinations of flat plate concentrators with central hole – fillet and central hole – groove.
- Definition of loads and the necessary supports to ensure the static equilibrium of the element.
- Running simulations and determination of equivalent Von Mises maximum stress present.

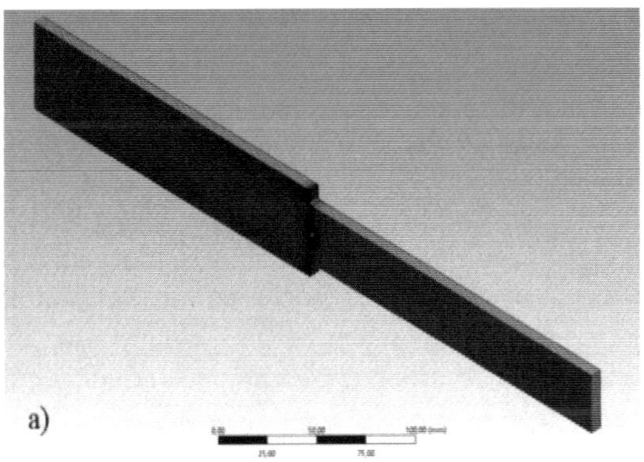

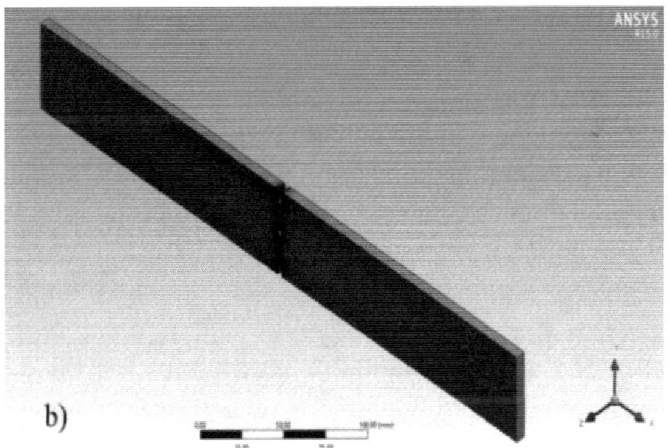

Figure 2. Result of meshing process for flat plate with (a) central hole – fillet and (b) central hole – groove.

3.2. Validation of the use of ANSYS software in obtaining stress concentration factors

The stress concentration factors were determined through Equation 1, by dividing the Von Mises equivalent stress obtained from simulations between the average stress determined through the fundamental equations of materials resistance to axial load. The use of ANSYS software in the determination of stress concentration factors is validated if, when you make a hypothesis test with the decision variable t of Student, no significant differences for a confidence level of 95% occur between the theoretical and simulated data.

3.3. Determination of K_c graphs on flat plates with combined concentrators

The simulations to determine stress concentration factors for static load conditions on flat plates with combined discontinuities (central hole – groove and central hole - fillet) were made to the following relations: wide (H/h) of 1.1 to 1.5 and 2.0; and radius (hole, fillet or groove) and smaller width (r/h) de 0.02, 0.05, 0.10, 0.15, 0.20, 0,25 and 0.30. The axial loads used in the simulations were 20000, 35000 and 50000N; these forces were applied specifically at the centroid of the cross section of the plate.

4. Obtained results

4.1. Validation of ANSYS software

The results of the maximum Von Mises stresses obtained from the simulations were used to determine the stress concentration factors in flat plates with a central hole and subjected to axial load; these data are listed in Figure 3.

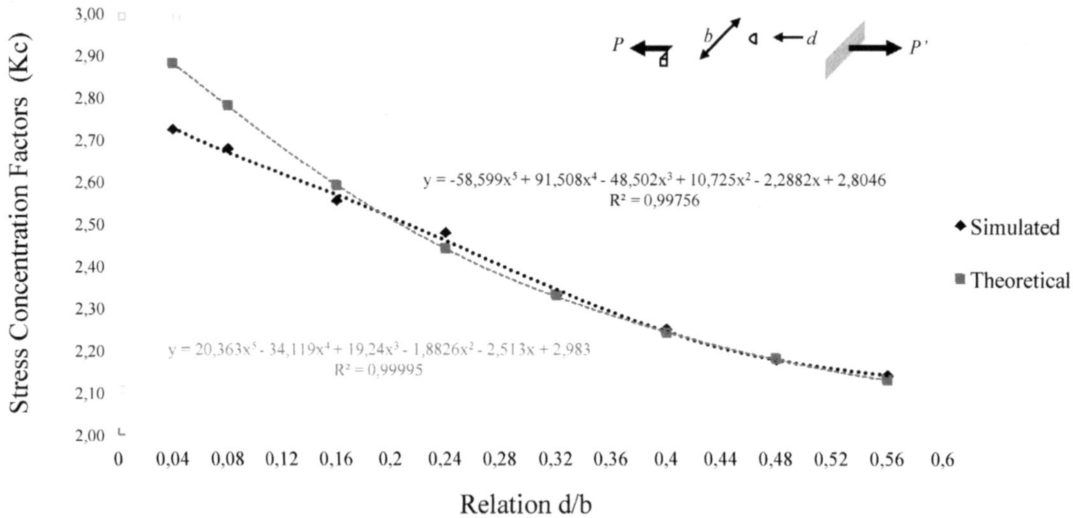

Figure 3. Comparison between stress concentration factor K_c theoretical and simulated, for a flat plate with a central hole.

To validate the use of ANSYS software in obtaining concentration factors, it was proceeded to determine the error % between the theoretical data and those obtained through simulation; in addition a statistical analysis provides an experimental decision variable of t Student of 0.237 value which is below the 2.1448 obtained from the table of this variable for a confidence of 95%, which allows to affirm that there are no significant differences between the theoretical and simulated data.

4.2. K_c stress concentration factor for flat plates with combined discontinuities

In Figures 4 and 5 can be observed the behaviour of stress concentration factors for flat plates with central hole – groove and central hole – fillet respectively; in both cases it is observed a polynomial relation of 5 degrees between Kc and the relation radius/smaller width (r/h).

In the combined discontinuity central hole – groove is observed that the concentration factor decreases as the radius of the discontinuities (hole and groove) increases, consistent with the behaviour exhibited in most concentration factors with simple discontinuities; additionally it can be concluded that there is no definite trend of this parameter with respect to the relation between widths (H/h) of the flat plate, which is due to the maximum value of stress is placed in some cases in the central hole and others in the groove.

In the combined discontinuity central hole – fillet is observed that the concentration factor decreases as the radius of discontinuities (hole and fillet) increases, in almost all the relations of (H/h), as for the r/h values lower than 0.05 in H/h equal to 1 has an opposite behaviour. Additionally it is concluded that there is no definite trend of this parameter with respect to the ratio between widths (H/h) of the flat plate, which is due to the maximum value of stress is placed in some cases in the central hole and others in the fillet.

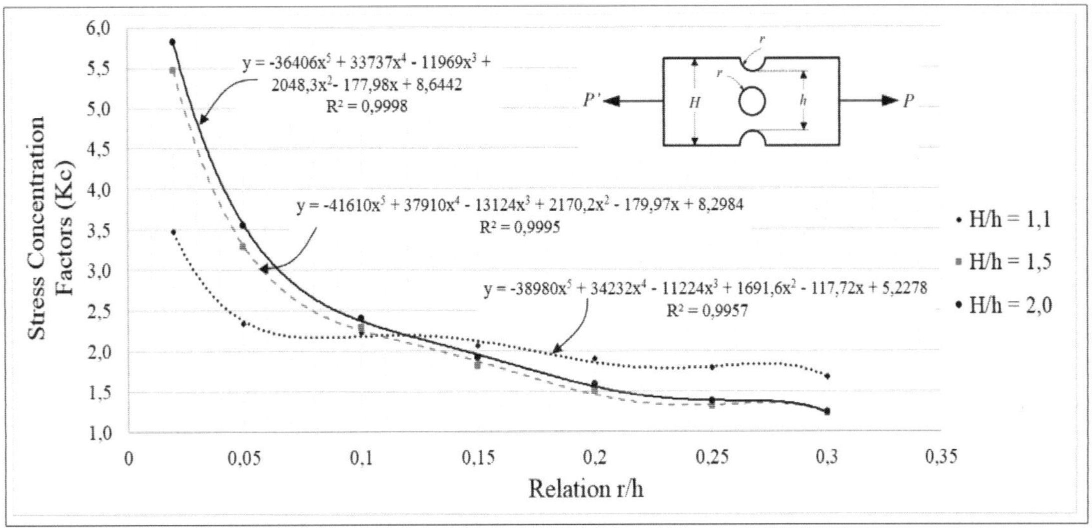

Figure 4. Stress concentration factor in a flat plate with central hole – Groove for axial load.

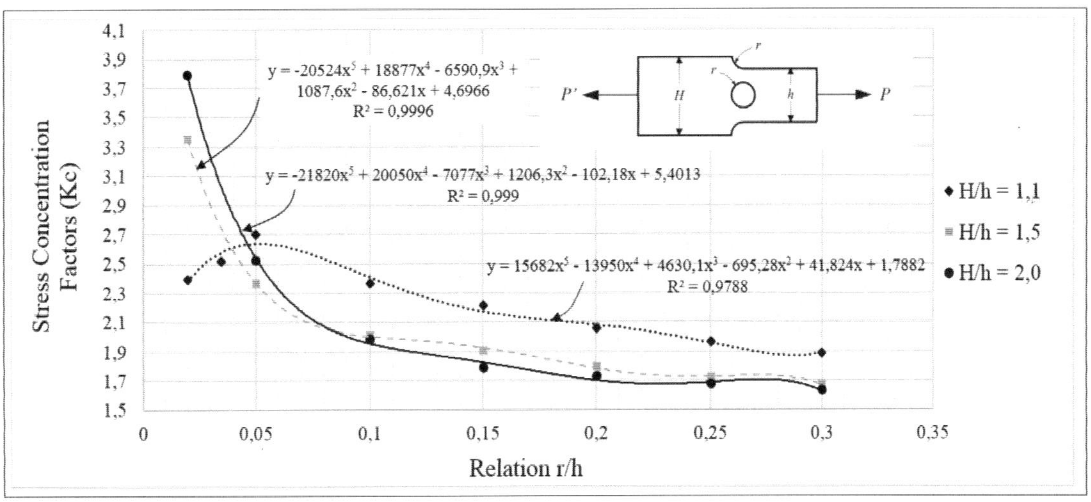

Figure 5. Stress concentration factor in flat plate with central hole – fillet for axial load.

5. Conclusions

The finite elements software can be used as a tool to determine graphs of stress concentration factors as for flat plates with a central hole and subjected to axial load, it can be statistically demonstrated that there are no significant differences between the theoretical data and the obtained data through simulation. The maximum error calculated between these data was 5.42%, when the geometrical relation between the hole diameter and width of sheet is 0.04.

In flat plates with a central hole – Groove is observed that the concentration factor decreases as the radius of discontinuities increases and there is no definite trend of this parameter with respect to the relation between widths (H/h) of the flat plate, which is due to the maximum value of stress in some cases is located in the central hole and others in the Groove.

In flat plates with central hole – fillet is observed that the concentration factor decreases as the radius of discontinuities increases in almost all relations (H/h), as the values r/h smaller than 0.05 in H/h equal to 1 has an opposite behaviour. Additionally it is concluded that there is no definite trend of this parameter with respect to the relation between widths (H/h) of the flat plate.

References

[1] R W Zimmerman 1988 *Mechanics Research Communications* **15** 87
[2] F Roldan and U Bastidas 2002 *Dyna* **69** 1
[3] R Amé, G M Dasso and D H Lezama 2011 *Revista cubana de ingeniería* **2** 43
[4] W Pilkey and D F Pilkey 2008 *Peterson's Stress Concentration Factors* (New Jersey: John Wiley & Sons) pp 6-24
[5] K Ting, K T Chen and W S Yang 1999 *International Journal of Solids and Structures* **36** 4
[6] A Santos 2013 *Journal of physics Conference Series* **466** 012035
[7] M F Spotts and T E Shoup 1999 *Elementos de Máquinas* (México: Prentice Hall) pp 128–142
[8] R G Budynas and J K Nisbett 2008 *Diseño en Ingeniería Mecánica de Shigley* (México: McGraw-Hill) p 210
[9] B J Hamrock, B O Jacobson and S R Schmid 2000 *Elementos de Máquinas* (México: McGraw-Hill) p 221
[10] J M Gere 2006 *Mecánica de Materiales 6a ed* (Australia: Thomson) pp 7–14
[11] T R Chandrupatla and A D Belegundu 2002 *Introduction to Finite Elements in Engineering 3rd edition* (New Jersey: Pearson Education) p 453
[12] O C Zienkiewics 1982 *El Método de los Elementos Finitos* (Barcelona: Reverté) p 903

Measurement temperature increment of open-celled cellular Zn-22Al-2Cu alloy

R Guzmán[1] and A Santos[1]
[1] Universidad Pontificia Bolivariana Seccional Bucaramanga, Colombia.

E-mail: rolando.guzman@upb.edu.co

Abstract. Thermo-mechanical properties of cellular materials, "metallic foams" make them very attractive in a variety of engineering applications. During plastic deformation of closed-cell metallic foams, part of plastic work is converted into heat. The generated heat increases may be quantified using an infrared camera measuring radiation emitted on the surface of the metallic foam. Experimental tests were carried out under quasi-static loading conditions using Zn-22Al-2Cu (zinalco) foams featuring different pore sizes, and densities between 30% to 50% maintaining constant volume. The goal of this study was to analyse the effect of heat generation during quasistatic compression of metallic foams at constant strain rate. Some conclusions on the mechanical behaviour were obtained in terms of temperature increase, the nominal stress-strain curves and relative density.

1. Introduction

The cellular materials offer excellent attractive combinations of mechanical properties [1-5]. Therefore, these materials have a potential of use in various structural applications, mainly in the automotive, aerospace and aeronautic industries [6-11]. The growing interest in the use of these materials is due to their high stiffness rigidity for a certain density and also due to their high capacity of absorbing energy under low stress values [6]. As a result of this absorption of energy, they are used more, in applications such as the protection of the interior space of the airship against impact. The cellular materials require a deep comprehension of their thermo-mechanical behaviour under static and dynamic conditions. During deformation of a solid, a part of mechanical energy is stored as internal defects, a part is dissipated for micro-structural changes as dislocations or phase transformation and the rest is converted into heat [12-17]. The last part changes the temperature field of the material loaded. This phenomenon depends of the material studied, the initial temperature, and the strain rate used during the test.

2. Experimental procedure

The test devices consist of three equipments: testing machine, infrared camera, and dark chamber [17].

2.1. Materials

Alloy Zn-22Al-2Cu is an eutectoid Zn-Al based with copper addition. Its thermo mechanical properties, as a solid, are listed on Table 1 [18].

Micrographs of the Zn-22Al-2Cu (zinalco) foams used are show in Figure 1. All specimens had cubic shape geometry, each of dimensions 1.5cmx1.5cmx3.5cm with different pore sizes, and densities between 30% to 50% maintaining constant volume. Table 2 shows properties of tested

Content from this work may be used under the terms of the Creative Commons Attribution 3.0 licence. Any further distribution of this work must maintain attribution to the author(s) and the title of the work, journal citation and DOI.
Published under licence by IOP Publishing Ltd

coupons. Bulk density was calculated measuring the mass of each coupon then dividing it by the known mass of a block made out of the solid material.

Table 1. Mechanical and thermal properties of the Zn-22Al-2Cu alloy.

Material	α (K^{-1})	K(W/mK)	Cp (J/KgK)	E(Gpa)	Tm(K)	ρ (Kg/m^3)
Zn-22Al-2Cu	$25x10^{-6}$	125.5	1000	90	694-754	$5.3x10^{-3}$

Figure 1. Micrographs of cellular materials (metallic foams).

Table 2. Properties of tested materials.

Zn-22Al-2Cu Sample	Mass (Kg)	Vol free (cm^3)	Bulk density (gr/cm^3)	Relative density (ρ^*/ρ_s)
N°1	13.111	5.447	1.665	0.300
N°2	14.563	5.178	1.848	0.342
N°3	14.813	5.132	1.881	0.348
N°4	15.457	5.013	1.963	0.363
N°5	16.154	4.883	2.051	0.379
N°6	16.141	4.886	2.050	0.379
N°7	16.007	4.911	2.033	0.370
N°8	17.441	4.645	2.215	0.410
N°9	18.665	3.456	2.370	0.438
N°10	19.124	4.333	2.428	0.449
N°11	20.098	4.153	2.552	0.472
N°12	21.387	3.914	2.716	0.500

2.2. Uniaxial compression

For quasi-static tests, a universal testing machine was used. The specimen has been deformed at a strain rate of $0.01s^{-1}$, and was carried out to a total deformation of 70%. The temperature evolution $\Delta T(t)$ and the curve stress-strain, $\sigma_n(\varepsilon_n)$ was measured during the test. The window size defined during the test to measure the temperature was 320x256 pixels with a frequency acquisition of 400FPS. The elastic behaviour of material tested is characterized by a stiffness parameter determined from the unloading portion an unload-load cycle in the test (3% nominal strain). The end surfaces were lubricated (MoS_2) to reduce the frictional restraint.

2.3. Measurement method

The temperature increase of the tested specimen was measured by a infrared camera (IRC). In a radiometrically calibrated IRC, the camera measurement has a linear relationship with the incoming radiance, $L_{in} = \int L_{in}(\lambda, T)d\lambda$ in a spectral range in which the camera is sensitive.

$$L_{in} = \sum_{i=0}^{n} C_i Measurement^i \quad (1)$$

The polynomial equation can be generated that converts the measurement "counts" to radiance and vice-versa (see Equation (1)). Figure 2(a) show the experimental values and the best adjusted fit, assuming linear detector response. For the IRC calibration, different images of a black body at different temperatures (0°C-125°C) (Figure 2(b)). In order to determine the temperature of the object, $L_{out}=L_{in}$ (total out radiance = total incoming radiance in the system sensor) is obtained from the measurement "counts". Knowing T_{amb} and the objects emissivity, L_0 is obtained. The object temperature is the value of T at which the integral of $L_0(\lambda,T)$ over the camera's spectral range equals L_0. The amount of spectral radiance $L_0(\lambda,T)$ emitted at a particular wavelength, λ, is given by Planck law (see Equation (2)), which states that for a black body:

$$L_0(\lambda, T) = \frac{C_1 \cdot \lambda^{-5}}{\exp\left(\frac{C_2}{\lambda T}\right) - 1} \quad (2)$$

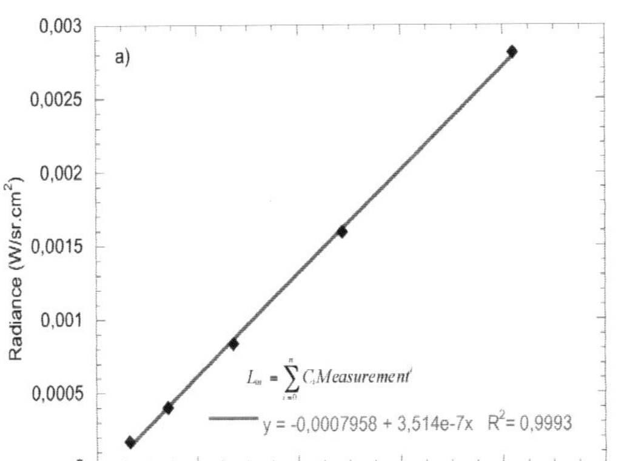

(a)

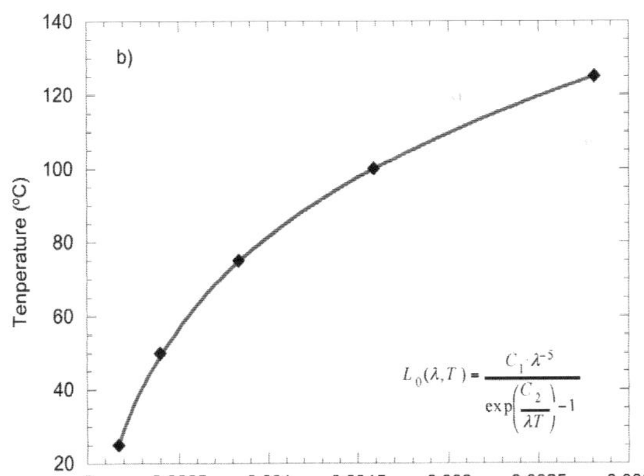

(b)

Figure 2: (a) Radiance vs Counts. (b) T vs Radiance; the solid lines are calculation with analytic integrate the Planck law.

The test devices consist of three equipments: testing machine, infrared camera, and dark chamber. Figure 3(a) and Figure 3(b) shows the set up experimental, identifying the specimen, the dark chamber, fixed head and cross head of the universal testing machine (Instron, model 8516).

In the IR remote sensing scenario, shown in Figure 3, the radiance emitted by the atmosphere and the IR system´s intrinsic radiation is considered negligible. Temperature increases are obtained on different pixels located on the surface of "cellular material".

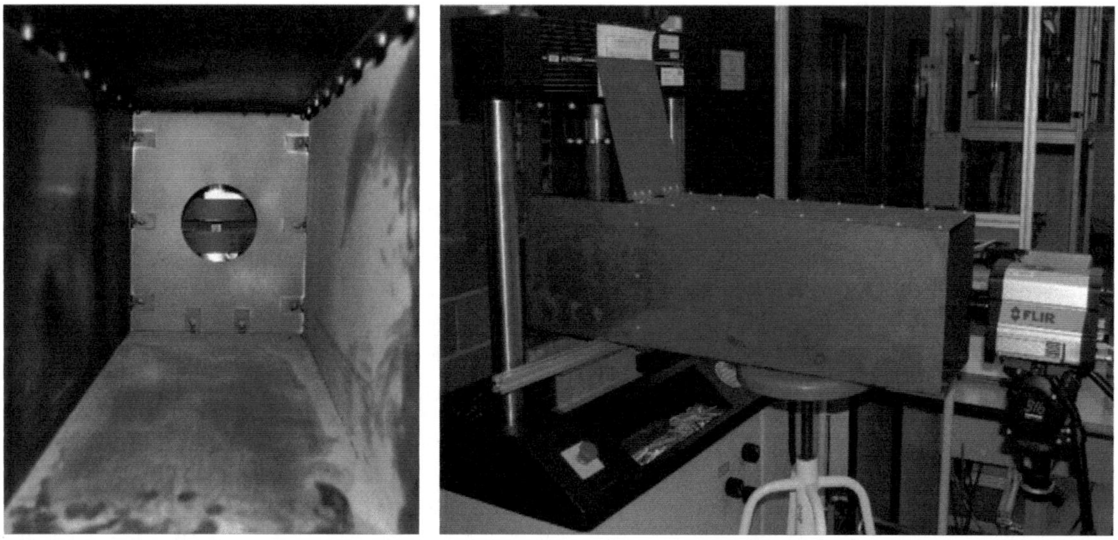

(a) (b)

Figure 3. (a) Dark chamber. (b) Infrared camera and dark chamber.

3. Results

Images shown in Figure 4 display a shear failure at 45° observed during the tests. Poruos walls exhibit failure when reaching maximum compressive strength. Tension drops significatively due to the sudden colapse of one of more high density layers. Figure 5 shows, as for example, the nominal stress-strain curves, obtained at room temperature for a strain rate 0.01s^{-1} (286K) with a relative density $\rho^* / \rho_s = 0.379$.

Figure 4. Failure mode of the specimen tested in compression for a strain rate of 0.01s^{-1}.

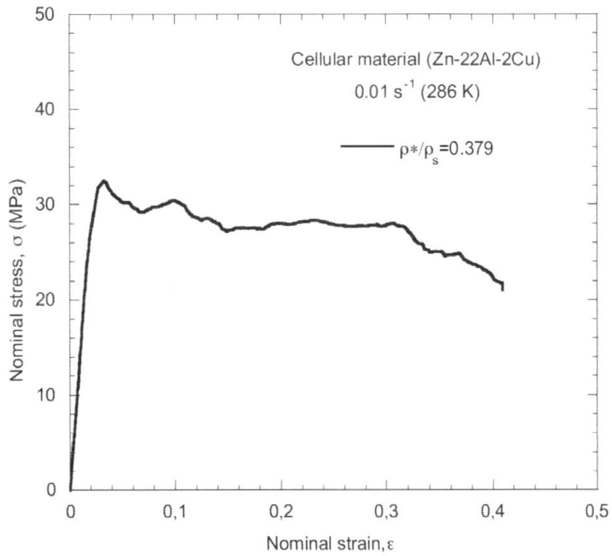

Figure 5. Nominal stress-strain curve for Zn-22Al-2Cu foam for a strain rate of 0.01 s^{-1} (286 K).

Measurement of incremental temperature (thermal field) associated with plastic strain variations were estimated along one line on the generatrix of the sample. An increasing temperature on the specimen surface was detected during the tests until total failure of the specimen took place. In addition, a rapid temperature increase was recorded just before the failure event and a process of temperature diffusion, was noted just after failure. Figure 6 shows the temperature evolution recorded by IRC technique for a strain rate of $0.01s^{-1}$ and a room temperature. The thermal field in Figure 6 shows a strain plastic localization (shear band).

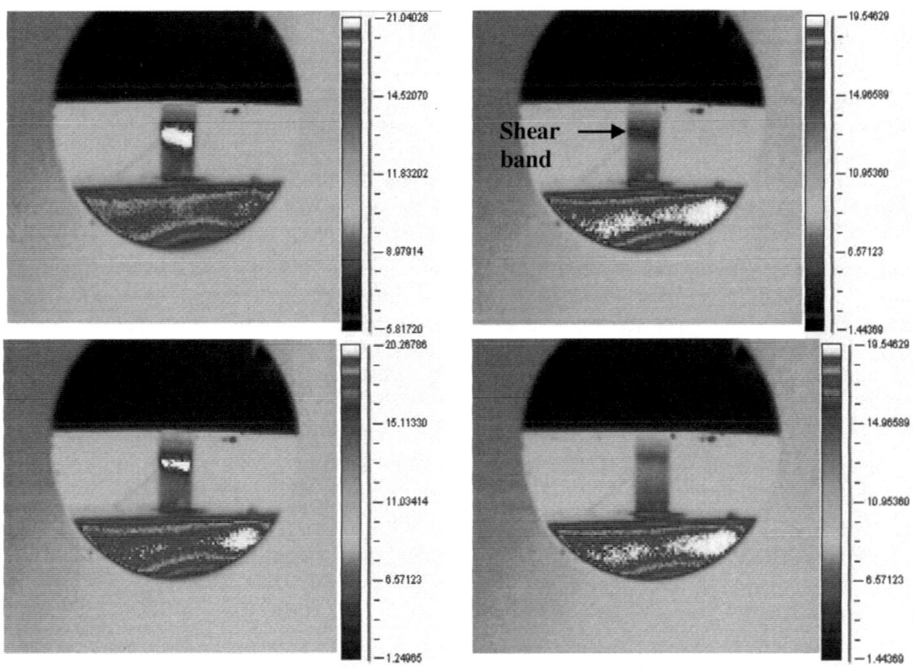

Figure 6. Images IRC for a test strain rate of $0.01s^{-1}$ and a room temperature.

4. Conclusions

The nominal stress-strain curves show three distinct regions: a linear elastic region at a low strain, a plastic plateau region with stress fluctuation over a wide range of the strain and a third region called the densification region, before fracture. For compressive quasi-static tests for cellular Zn-22Al-2Cu foams, an increment in relative density raises the temperature increment associated to plastic strain. This is because an increase in relative density raises the deformation energy absorbing capacity by the foam.

Both in load-time and in temperature-time relations, oscillations were detected. The appearance of these oscillations can be associated to the wall´s progressive fail in Zn-22Al-2Cu foam. Less porous walls fail when reaching a maximum compressive strength. The tension drop shown is due to the sudden fracture of high density walls.

Pore size affects alloy Zn-22Al-2Cu during compressive strengh because smaller pores show a more stable stress flow hence, showing a flat zone in the σ-ε curve (Plateu stress). This is reflected in higher energy absorption values. Therefore, it can be concluded that, for energy absorption, for Zn-22Al-2Cu foams with a high density is more effective than the one with a lower density.

Acknowledgements

The authors would like to express their gratitude to Engineering Department of Universidad Carlos III de Madrid and Universidad Nacional Autónoma de Mexico.

References

[1] Banhart J 2000 *Progress in Materials Science* **46(6)** 559-632
[2] Banhart J and Baumeister J 1998 *Journal of Materials Science* **33(6)** 1431-1440
[3] Gama B, Bogetti T, Fink B, Yu C, Claar T, Eifert H, Gillespie J 2001 *Composite Structures* **52(3-4)** 381-395
[4] Hahfuz H, Zhu Y, Haque A, Abutalib A, Vaidya U, Jeelani S 2000 *International Journal Impact Engineering* **24(2)** 203-217
[5] Villanueva G and Cantwell W 2004 *Composite Science and Technology* **64(1)** 35-54
[6] Baumeister J, Banhart J, Weber M 1997 *Materials & Design* **18(4-6)** 217-220
[7] Fuganti A, Lorenzi L, Grønsund A, Langseth M 2000 *Advanced Engineering Materials* **2(4)** 200-204
[8] Stöbener K, Baumeister J, Lehmhus D, Stanzick H, Zöllmer V 2003 Composites based on metallic foams: Phenomenology, production, properties and principles *Advanced Metallic Materials: Proceedings of the International Conference* ed J. Jerz (Slovakia: Institute of Materials & Machine Mechanic) pp 281-286
[9] T. Bernard, H.W. Bergmann, C. Haberling and H. G. Haldenwanger 2002 *Advanced Engineering Materials* **4(10)** 798-802
[10] Yu C J, Eifert H H, Banhart J, Baumeister J 1998 *Materials Research Innovations* **2(3)** 181-188
[11] Seeliger H W 2002 *Advanced Engineering Materials* **4(10)** 753-758
[12] Rusinek A and Klepaczko JR 2009 *Materials & Design* **30(1)** 35-48
[13] Rusinek A, Gadaj P, Klepaczko J R and Nowacki W K 2004 *Matériaux & Techniques* **92(5-6)** 21-30
[14] Rusinek A, Novacki W K, Gadaj P, Klepaczko J R 2003 *Journal de Physique IV* **110** 411-416
[15] Hodowany J, Ravichandran G, Rosakis A J, Rosakis P 2000 *Experimental Mechanics* **40(2)** 113-123
[16] Netzelmann U, Abuhamad M and Walle G 2005 *Journal de Physique IV* **125** 511-514
[17] Guzmán R, Essa Y, Meléndez J, Aranda J, López F, Pérez-Castellanos J L 2009 *Strain* **45(2)** 179-189
[18] Elizabeth Martínez-Flores and Gabriel Torres-Villaseñor 2011 Hybrid Materials Based on Zn-Al Alloys *Metal, Ceramic and Polymeric Composites for Various Uses Edition 1st* ed John Cuppoletti (Croatia: Intech open science) Chapter 7 pp 149-170.

Estimate of the concentration of implanted ions in solid substrates using a web application

F H Vera Rivera[1,2], B R Pérez Gutiérrez[2], H J Dulce-Moreno[1,2], F Duran-Flórez[4] and E D V-Niño[1,3,5]

[1] Foundation of Researchers in Science and Technology of Materials, Colombia.
[2] Universidad Francisco de Paula Santander, San José de Cúcuta, Colombia.
[3] Universidad Industrial de Santander, Bucaramanga, Colombia.
[4] Universidad Pontificia Bolivariana, Bucaramanga, Colombia.
[5] Universidad Politécnica de Madrid, Madrid, España.

E-mail: fredyhumbertovera@ufps.edu.co

Abstract. The three-dimensional ionic implantation technique (3DII) is used to modify the surface of solid metal by electric discharges pulsed of high voltage at low pressures. Knowing the density of ions implanted in the surface of a functional element, in a faster and estimated way, will help to optimize the surface treatment technique. Therefore, a web application was developed which from experimental parameters established in a process 3DII estimates the concentration of ions implanted in solid metal substrates. The results obtained in this research work demonstrate the feasibility of the computational web tool to perfect the experiments of surface modification by ion implantation.

1. Introduction

The modification superficial consists in alter the composition and structure of materials either by creating a coating (with typical thicknesses of a few microns) or by introducing new elements into the surface (dopant ions) in nanometric depths [1].

The three-dimensional ion implantation (3DII) [2] is a superficial modification process generated by high voltage discharges at low pressures, which turns on in the left branch of the Paschen curve, and it is used as an alternative to improve physical and chemical properties of materials [3-6]. The pioneers in the study of this type of discharge were Klarfeld and Pokrovskaya-Soboleva [7] and McClure [8]. The reactor JUPITER (Join Universal of Plasma and Ion Technologies Experimental Reactor) [9] is the only equipment built based on this technology that allow the implantation of ions in solid substrates of metallic and non-metallic species. The parameters that characterize the implementation are: the ion's type, the ion's energy and the implanted dose [1, 2, 10-12].

The calculation of superficial density of ions implanted in some material depends mainly of time of treatment, density of ionic current, repetition frequency and pulse duration, the secondary emission ion-electron coefficient of material and the area of the cathode. This paper presents a computational tool, which allows us estimate the value for implanted dose on the surface of solid substrates after a process of surface modification by ion 3DII. These calculations are faster and correct that traditional form. The web application is the result of a technological applied research, which can be used at any time, from anywhere, in different electronic devices (computers, tablets, mobile phones, etc.).

Content from this work may be used under the terms of the Creative Commons Attribution 3.0 licence. Any further distribution of this work must maintain attribution to the author(s) and the title of the work, journal citation and DOI.
Published under licence by IOP Publishing Ltd

2. Methodology

The estimation of superficial implanted dose of a material using the web application is carried out according to the scheme presented in Figure 1.

In first place the experimental data are acquired from the JUPITER reactor, this data correspond to the data and image of the voltage and current pulses of electrical discharge used in the ion implantation process. These data and image are introduced into the web application for subsequent scanning and digitalization; after the graphics are displayed, next the user proceeds to set the pulse width depending on the region of interest, then the total electrical charge is estimated using the method proposed in [12]. Finally, the superficial implanted dose of solid substrates is estimated.

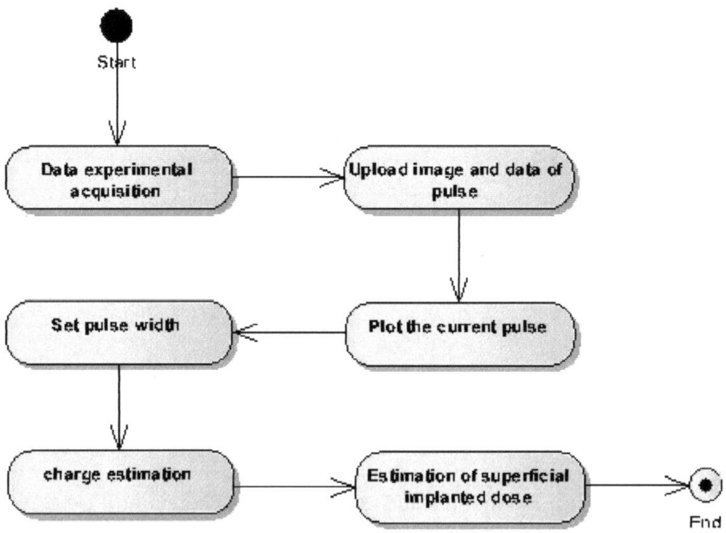

Figure 1. Methodology of estimation superficial implanted dose.

3. Results - web application

The algorithm for the estimation of surface ion implantation was developed in JAVA, the web application was implemented in Java Server Pages (JSP) using Google Chart to plot and digitize the pulse of the implementation process.

The process begins with the acquisition of data and image pulse, after the user upload the information into the web app, he has the possibility of set the pulse width and provide the input data (experimental parameters of the ion implantation process) which are: frequency (f) of the discharge pulse (Hz), total time (t) of the discharge (s), total area (A) exposed (cm^2), number of ion of the charge (Ze) and the secondary emission coefficient (γ). The web application calculates the total pulse charge (Q) [4] and the dose approximate of ions implanted superficially using the Equations (1)-(6).

$$n = f * t \tag{1}$$

$$Q_T = Q * n \tag{2}$$

$$Q_T = Q_i + Q_e \tag{3}$$

$$Q_e = \gamma * Q_i \tag{4}$$

$$Q_i = \frac{Q_T}{\gamma + 1} \tag{5}$$

$$D = \frac{Q_i}{A * Ze} \tag{6}$$

Where, n: pulses number. Q_T: total charge transported, Q_i: ionic charge, Q_e: electronic charge.

Next in Figures 2-5 are presented in detail how the calculation of the ions implanted superficially through the web application is performed. Figure 2 shows how to load the image and data pulse will be digitized.

Figure 2. Upload pulse data and image.

In Figure 3, the pulse's image obtained from the reactor JUPITER is presented, the web application shows the graphic of the digitized pulse. The user proceeds to define the pulse width and establish the experimental parameters required in the surface modification process (see Figure 4).

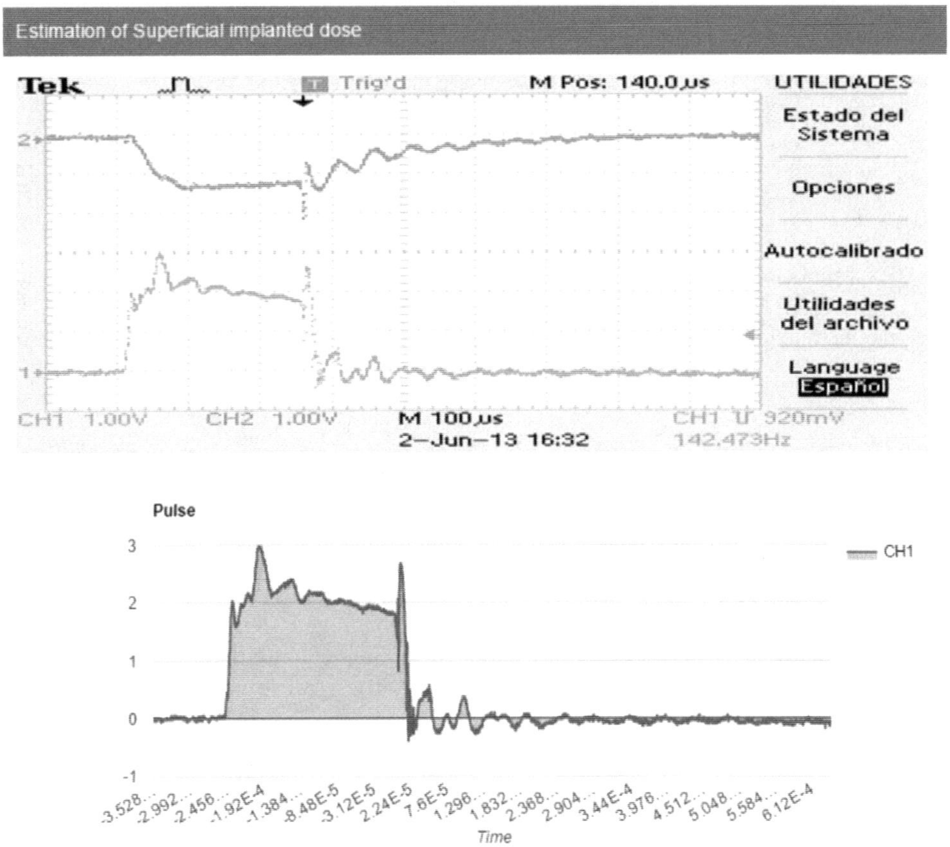

Figure 3. Current pulse plot.

Journal of Physics: Conference Series **743** (2016) 012016 doi:10.1088/1742-6596/743/1/012016

Figure 4. Experimental parameters of the ion implantation process.

Finally, the web application proceeds to estimate the superficial implanted dose according to Equations (1) to (6) and whose value is reported in Figure 5.

Figure 5. Estimation of superficial implanted dose.

4. Conclusions

This paper presents a web application that provides a friendly and easy-to-use tool to estimate the dose of ion concentration implanted in solids. These calculations are faster and more precise than traditional methods.

The construction of computational tools to speed up and improve the calculation of this type of process brings benefits to the Plasma Laboratory. Researchers can focus on other tasks and not spend time making complex calculations which can lead them to inaccurate and faulty implementations.

The calculations for the estimation of the surface dose of ions obtained from the web application are very similar to the theoretical calculations, ensuring that the implemented web application works correctly.

References

[1] Dougar-Jabon V D, Dulce Moreno H J and Tsygankov P A 2003 Implantación Iónica *Revista Respuestas* **8** 19-30
[2] Khvesyuk V I and Tsygankov P A 1997 The use of a high-voltage discharge at low pressure for 3D ion implantation *Surf Coat Technol* **96(1)** 68-74
[3] Dougar-Jabon V D, Dulce-Moreno J and Tsygankov P A 2002 High voltage pulse discharge for ion treatment of metals *Rev Sci Instrum* **73(2)** 828-830
[4] Dulce-Moreno H J, Tsygankov P A, Dugar-Zhabon V, V Niño E D and Parada F F 2011 Estudio de los haces electrónicos en la descarga eléctrica de alto voltaje a bajas presiones *Puente Revista Científica* **5(2)** 21-26
[5] Parada-Becerra F, Cabanzo R, Dugar-Zhabon V, Tsygankov P, Mejia-Ospino E and V Niño E D 2012 Plasma temperature measurement in a hybrid discharge by using optical diagnostics *Journal of Physics Conference Series* **370** 012054
[6] V Niño E D, Dugar-Zhabon V, Dulce-Moreno J, Peña-Rodríguez G, Garnica H A and Tsygankov P 2012 Aplicación de descargas simultáneas de alto voltaje y arco eléctrico para el tratamiento superficial avanzado de metales *Iteckne* **9(1)** 14-20
[7] Pokrovskaya-Soboleva A S and Klarfeld B N 1957 Ignition of the High Voltage Discharge in Hydrogen at Low Pressures *Sov Phys JETP* **5(5)** 812-818
[8] McClure G W 1961 High-Voltage Glow Discharges in D_2 Gas. I. Diagnostic Measurements *Phys Rev* **124(4)** 969-982
[9] Dougar-Jabon V D, Castro B J, Dulce Moreno H J, Tsygankov P A 1999 Device JUPITER for ion implantation *Revista Colombiana de Física* **31(2)** 181
[10] V Niño E D and Dougar-Jabon V D 2006 Comportamiento del acero SAE-4140 implantado con iones de nitrógeno en ambientes hidrogenados *Revista Colombiana de Física* **38(1)** 61-64
[11] Valbuena Niño E D, Dulcé M H J and Dugar-Zhabon V 2010 Caracterización del acero AISI 4140 implantado por iones de nitrógeno *Revista Colombiana de Física* **42(3)** 387-392
[12] F H Vera-Rivera, V D Dugar-Zhabon, H J Dulce-Moreno and E D V-Niño 2016 Development of a web application for estimate the total charge in an electric discharge *Journal of Physics Conference Series* **687** 012058

Estimation of the ionic charge of non-metallic species into an electrical discharge through a web application

B R Pérez Gutiérrez[2], F H Vera-Rivera[1,2] and E D V-Niño[1,3,4]
[1] Foundation of Researchers in Science and Technology of Materials, Colombia.
[2] Universidad Francisco de Paula Santander, San José de Cúcuta, Colombia.
[3] Universidad Industrial de Santander, Bucaramanga, Colombia.
[4] Universidad Politécnica de Madrid, Madrid, España.

E-mail: borisperezg@ufps.edu.co

Abstract. Estimate the ionic charge generated in electrical discharges will allow us to know more accurately the concentration of ions implanted on the surfaces of nonmetallic solids. For this reason, in this research a web application was developed to allow us to calculate the ionic charge generated in an electrical discharge from the experimental parameters established in an ion implantation process performed in the JUPITER (Joint Universal Plasma and Ion Technologies Experimental Reactor) reactor. The estimated value of the ionic charge will be determined from data acquired on an oscilloscope, during startup and shutdown of electrical discharge, which will then be analyzed and processed. The study will provide best developments with regard to the application of ion implantation in various industrial sectors.

1. Introduction

The three-Dimensional Ion Implantation technique (3DII) [1] performed in the JUPITER reactor [2] is a process of introducing dopant ions from nonmetallic species into metallic solid materials [3-6]. Therefore, change (no change in geometric dimensions) surface and sub-surface layers of solid substrates are subject of great interest in the energy sector, transport and manufacturing [7-10].

The 3DII generated from hybrid discharges of high voltage and electric arc at low pressures, is to make an impact with sufficient energy (between 5 and 60 kV) ions of non-metallic species on the surface of a solid substrate, which penetrate a few nanometric layers making change physical and chemical properties of the surface [11-12]. During travel of the ion into the material, a lot of interaction (collision) with the particles (atoms) composing the solid substrate are produced, which cause a series of collective processes which influence the final distribution of ions, such as the formation of defects, ions tunneling, cascade collisions, sputtering, etc. [13-17].

According to the above and considering that the final concentration of implanted ions is a very important feature of the surface modification by ion implantation [18-20], it is very important to characterize the estimation of the ionic charge in metallic and non-metallic species, moreover, there is a need to understand the complex phenomena that occur during the process. Therefore, the development of a web application with a computer simulation are both really important, especially because we can obtain information that today could not be obtained by other means. That is why it is necessary to develop computational tools that estimate more quickly, accurately and precisely the ionic charge generated in an electric discharge [21].

Content from this work may be used under the terms of the Creative Commons Attribution 3.0 licence. Any further distribution of this work must maintain attribution to the author(s) and the title of the work, journal citation and DOI.
Published under licence by IOP Publishing Ltd

In this paper, we present EIGE - Estimation of the Ionic Charge Generated in an Electrical Discharge, a free web application that provides an user-friendly tool to estimate the ionic charge generated in an electric discharge. EIGE was developed in a joint effort between the Software Engineering Development Research Group (GIDIS) of the Universidad Francisco de Paula Santander and the Physics and Plasma Technology and Corrosion Research Group (FITEK) of the Universidad Industrial de Santander.

2. Experimental

In this article we use the methodological model proposed in [21] where the experimentation was divided in three phases. These phases allow us to retrieve the required data, processes it using an in-house web application and finally, display the results within the web application.

2.1. Data acquisition

The current pulses versus time during electric high voltage discharge at low pressures were obtained from the experimental parameters set in a process of nitrogen ion implantation into metallic solid substrates. The pulses recorded represent the total electronic charge transported.

2.2. Design and development of web application

A web application is a client-server application where the client runs in a web browser, like Chrome or Firefox. This is a web based application having two-tier architecture. Web application was developed using HTML, Google Charts, Java Server Pages and BootStrap. The latter help us to developing responsive web applications.

The design and development of this web application was made to quickly and accurately calculate the ionic charge generated in electrical discharges from the data acquired of a set of current pulses. Two data sets were required to be uploaded to the web app. Next, the obtained data was digitized and graphed in order to present it to the user. After that, the area under the curve was calculated in both cases, and then their difference was obtained.

2.3. Output data

The estimated data of the resulting area under curve (estimated total ionic charge) is reported through the web application.

3. Web application

A web application was implemented to quickly and accurately calculate the ionic charge from a pulse of electric current and the total electrical charge, both in function of the time. Figure 1 shows the use case diagram describing the available features for the user on the web application.

The web application allows the user to upload two images and two datasets in CVS format (see Figure 2). Each one of these datasets corresponds to the pulses of current that can be captured on an oscilloscope during turn on of the electric high voltage discharge (in this case we use the oscilloscope Tektronix TDS 2002 B).

Figure 1. Use case diagram showing the functionality of the system.

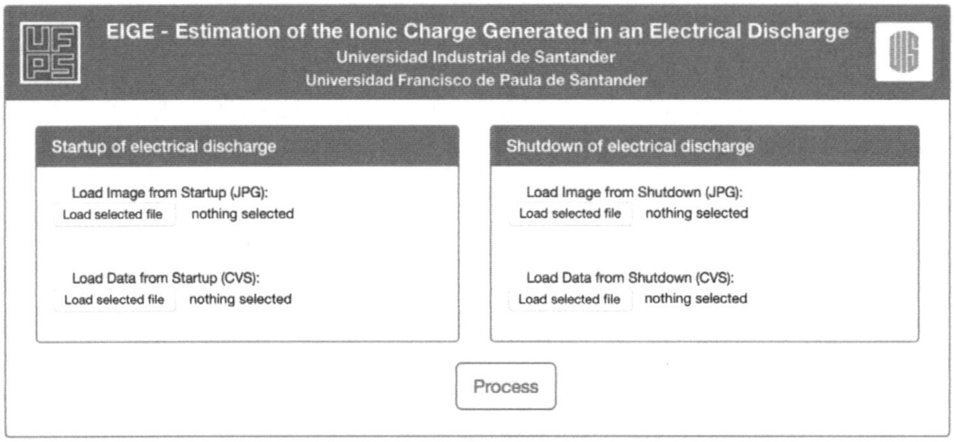

Figure 2. Screen to load the image and the pulse data.

After user loads the required data and clicked in Process button, the web app performs a process of digitization of datasets and graphics. The screen is splitted in two sides, each one presenting the graphic along with the image of the pulses. The web app allows the user to set the maximum and minimum values to determine the pulse width necessary to calculate the area under the curve and estimate the transported load for each pulse (see Figure 3).

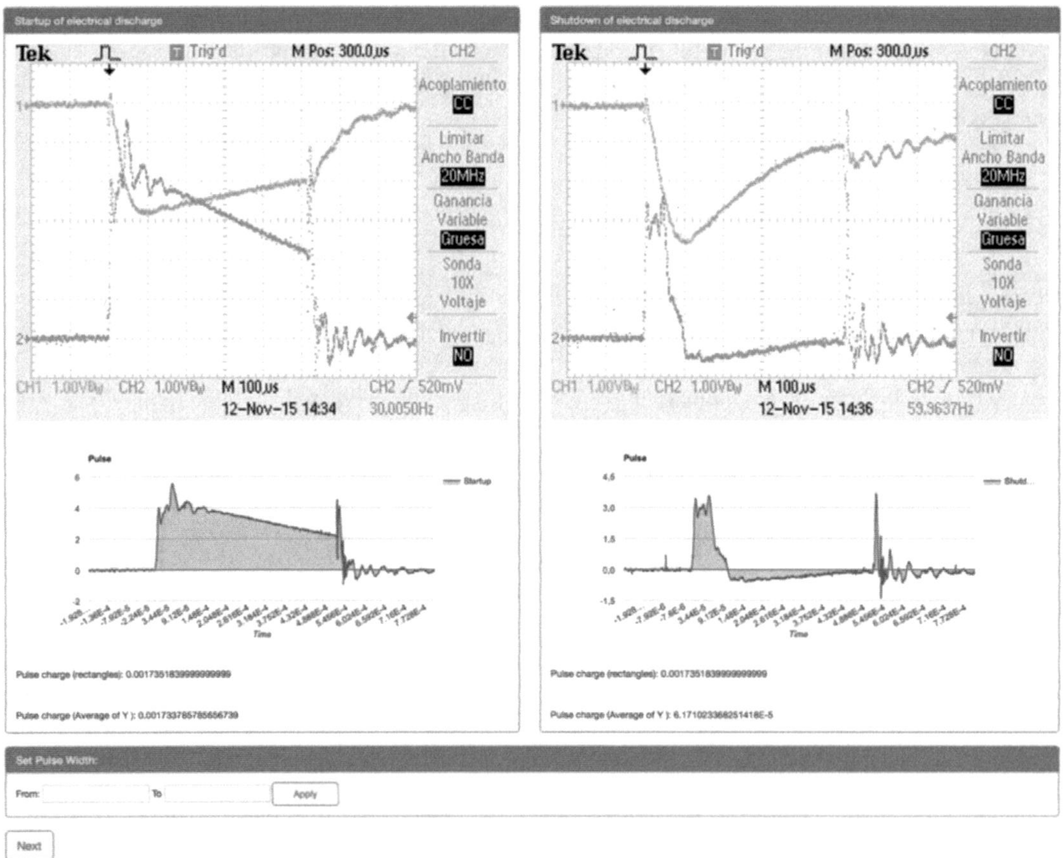

Figure 3. Images and graphics of pulses in the web application.

Journal of Physics: Conference Series **743** (2016) 012017 doi:10.1088/1742-6596/743/1/012017

On top of Figure 3 is shown both pulses images obtained from Tektronix TDS 2002 B oscilloscope. Left one corresponds to the total charge transported during the electric high voltage discharge. Right one corresponds to the electric charge carried out. At the bottom of Figure 3 both pulses are graphed in different charts using our web application (the graph obtained is very similar compared to the graph acquired with the oscilloscope). User can filter the pulse width. In this case, user enter the required data corresponding to the X axis (Figure 4).

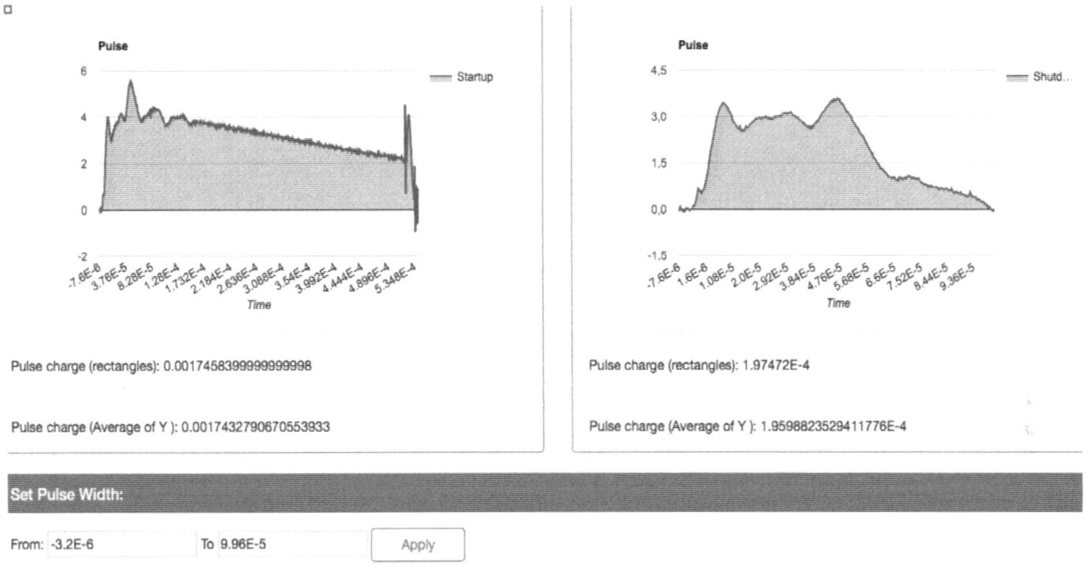

Figure 4. Pulse width adjusted by user.

Finally, user clicked the Next button to show both charts mixed in one chart, so the user can quickly identify the calculated area, corresponding to the ionic charge. Values entered in pulse width textbox are kept in this chart. Finally, the estimated value of the ionic charge is presented (Figure 5).

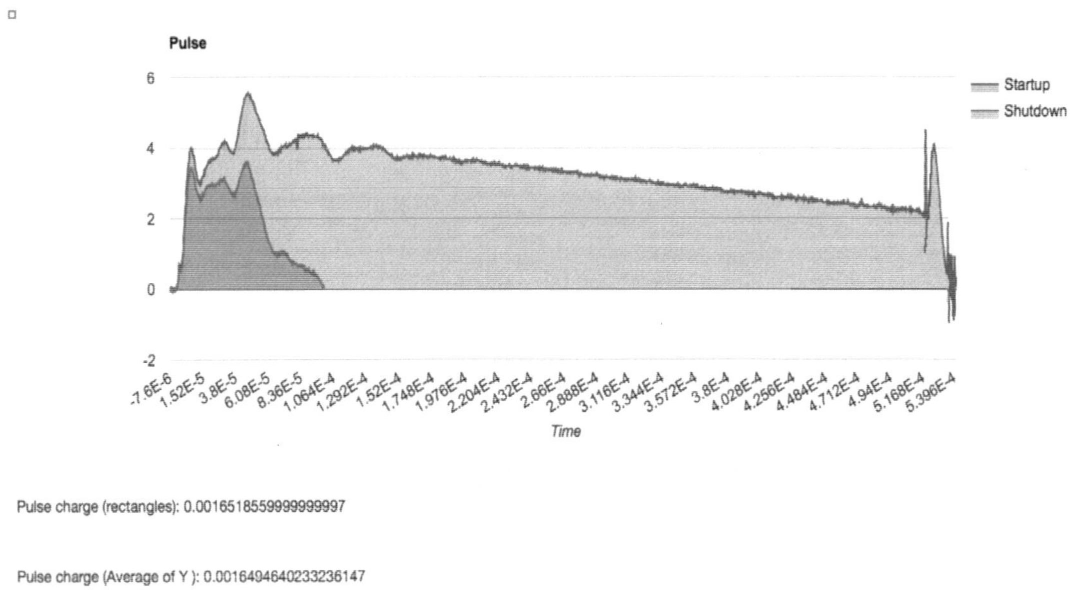

Figure 5. Results of the estimate ionic charge.

4. Conclusions

A computational web tool was built to streamline the estimation of the calculation of the ionic charge carried out in electrical discharges of high voltage at low pressures, bringing benefits to experimental researches in the UIS' plasma laboratory.

The web application developed can be used anywhere any time, only an Internet connection and the pulses data to be processed are required. The format of the data received by the system can be downloaded from the web application.

The processed data obtained from the web application are really similar to current pulses reported experimentally during the turn on of the electric discharge of high voltage, ensuring that the implemented web application works correctly to estimate the ionic charge.

References

[1] Dougar-Jabon V D, Castro B J, Dulce Moreno H J, Tsygankov P A 1999 *Rev Colom de Física* **31(2)** 181
[2] V I Khvesyuk, P A Tsygankov 1997 *Surface & Coatings Technology* **96** 68
[3] Dougar-Jabon V D, Dulce Moreno H J and Tsygankov P A 2003 *Revista Respuestas* **8** 19
[4] Conrad J R 1988 *Method and apparatus for plasma source ion implantation* (USA: Patent No 4764394)
[5] Conrad J R, Radtke J L, et al 1987 *J Appl Phys* **62** 4591
[6] J R Conrad 1987 *Mater Sci Eng* **116A** 197
[7] Ely Dannier V Niño, Hernán Garnica, Veleriy Dugar-Zhabon and Genis Castillo 2014 *Journal of Physics Conference Series* **511** 012080
[8] J J Castro-Maldonado, H J Dulcé-Moreno and E D V-Niño 2013 *Journal of Physics Conference Series* **466** 012005
[9] E D V-Niño, V Dugar-Zhabon, H A Garnica 2011 *Revista Iteckne* **8(1)** 37
[10] Héctor J Dulcé M, Alejandro Rueda V and Valeri Dougar-Jabon 2005 *Phys Stat Sol* (c) **2(10)** 3778
[11] H Jaime Dulce M, Valeri Dougar J and Carlos R Cabrera 2003 *Revista Colombiana de Física* **35(2)** 332
[12] V Dougar, J Dulcé and P A Tsygankov 2002 *Rev Sci Instr* **73** 828
[13] Edmanuel Torres, Valeri Dougar-Jabon 2006 *Revista Colombiana de Física* **38(1)** 89
[14] K Krantzman, Z Postawa, B Garrison, N Winnograd, S Stuart and J Harrison 2001 *Nucl Instr and Meth in Phys Res B* **180** 159
[15] K Nordlund, J Keinonen, M Ghaly and R S Averback 1999 *Nucl Instr Meth Phys Res B* **148** 74
[16] J F Ziegler, J P Biersack, U Littmark 1985 *The stopping and range of ions in matter* (New York: Pergamon)
[17] Yu V Gott 1978 *Interaction of Particles with substances in plasma research* (Moscow: Atomizdat)
[18] Towsend P D, Chandler P J and Zhang L 1994 *Optical effects of ion implantation* (Cambridge: Cambridge University press)
[19] Wolfgang Eckstein 1991 *Computer simulation of ion-solid interactions* (Berlin: Springer-Verlag)
[20] Burcham W E 1979 *Elements of nuclear physics* (New York: Longman Group Limited)
[21] F H Vera-Rivera, V D Dugar-Zhabon, H J Dulce-Moreno and E D V-Niño 2016 *Journal of Physics Conference Series* **687** 012058